EARTH TRANSFORMED

EARTH TRANSFORMED

William F. Ruddiman

W.H. Freeman and Company
A Macmillan Higher Education Company

To the many scientists who have assembled the story of how farming spread across the Earth.

Associate Publisher: Jessica Fiorillo
Senior Acquisitions Editor: Bill Minick
Development Editor: Tony Petrites
Senior Marketing Manager: Alicia Brady
Marketing Assistant: Alissa Nigro
Cover and Text Designer: Blake Logan
Project Editor: Kerry O'Shaughnessy
Production Manager: Paul W. Rohloff
Illustration Coordinator: Janice Donnola
Illustrations: Robert L. Smith
Photo Editor: Bianca Moscatelli
Photo Researcher: Julie Tesser
Composition: Aptara
Printing and Binding: RR Donnelley

The "About the Author" photo is courtesy of Ginger Ruddiman. Credit is also given to the following sources for the part opener photos:
Part 1 (p. 1): Corbis Premium RF/Alamy; **Part 2** (p. 43): FLPA/Wayne Hutchinson/age fotostock; **Part 3** (p. 135): Terry Whittaker/Alamy; **Part 4** (p. 209): Patrick Frilet/age fotostock; **Part 5** (p. 245): Pierre Zeni/Still Pictures/Robert Harding; **Part 6** (p. 287): Joel Sartore/National Geographic/Getty Images.

Library of Congress Control Number: 2013930524

ISBN-13: 978-1-4641-0776-4
ISBN-10: 1-4641-0776-9

W. H. Freeman and Company
41 Madison Avenue
New York, NY 10010
Houndmills, Basingstoke RG21 6XS, England
www.whfreeman.com

CONTENTS

About This Book   vii

About the Author   xi

Prologue   Did Civilization Develop in a Naturally Warm World?   xiii

PART 1   A Mystery: Wrong-Way Greenhouse-Gas Trends   1

Chapter 1   Nature's Climatic Cycles   5

Chapter 2   Wrong-Way Methane Trend   19

Chapter 3   Wrong-Way Carbon Dioxide Trend   31

PART 2   Early Agriculture: Answer to the CO_2 and CH_4 Mysteries?   43

Chapter 4   The Fertile Crescent and Europe   51

Chapter 5   China and Southern Asia   73

Chapter 6   The Americas   95

Chapter 7   Africa, Australia, and Oceania   117

PART 3   Debating a New Hypothesis   135

Chapter 8   Early Farming and Per Capita Land Use   139

Chapter 9   How Should Interglacial Gas Trends Be Compared?   159

Chapter 10   Natural Versus Anthropogenic CH_4 Sources: Closer Scrutiny   175

Chapter 11   Natural Versus Anthropogenic CO_2 Sources: Closer Scrutiny   189

PART 4   How Science Moves Forward   209

Chapter 12 Falsification   213

Chapter 13 Paradigm Shifts   221

Chapter 14 An Emerging Paradigm for the Anthropogenic Era?   231

PART 5   Early Human Effects on Climate   245

Chapter 15 Is the Next Glaciation Overdue?   249

Chapter 16 Other Climatic Effects of Early Land Clearance   263

Chapter 17 The End of Northern Hemisphere Glaciations   275

PART 6   Small Steps Back Toward an Ice Age   287

Chapter 18 The Little Ice Age   291

Chapter 19 Were the Drops in CO_2 and CH_4 Natural?   303

Chapter 20 Mass Human Mortality and CO_2 Decreases   311

Chapter 21 Effects of Humans on Short-Term Greenhouse-Gas Reductions   331

Epilogue   345

Glossary   353

Index   361

ABOUT THIS BOOK

This book differs from others designed for teaching science-based courses. Rather than first laying out the basics of a single discipline and then gradually building to a higher level of complexity, this book immediately jumps into a scientific debate that has been underway for more than a decade. The central issue is whether major human effects on Earth's environment and climate began around 150 years ago with the rise of greenhouse gases emitted during the Industrial Revolution, or instead thousands of years earlier as part of the Neolithic Agricultural Revolution following the discovery and spread of agriculture. The structure of this book is explained below.

Part 1 introduces a "mystery": the concentrations of two greenhouse gases—carbon dioxide (CO_2) and methane (CH_4)—measured in air bubbles in ice cores rose during the last several thousand years of the current warm interglacial climate, yet fell during equivalent times in three previous interglaciations (intervals when no ice sheets existed in the Northern Hemisphere except the one on Greenland). The seemingly "wrong-way" directions of the two gas trends during recent millennia suggest an origin new to the climate system.

Part 2 provides a possible answer to this mystery by documenting the spread of agriculture across Earth's major continents during the last 10,000 years. Early farming activities are known to have emitted greenhouse gases to the atmosphere, and the gradual spread of agriculture coincides with the anomalous CO_2 and CH_4 increases.

Part 3 traces the key issues in the decade-long debate over natural versus human explanations for the greenhouse-gas increases. The section arrives at the overall assessment that an agricultural origin for the wrong-way greenhouse-gas trends is plausible but that future archeological work mapping the spread of agriculture is needed.

Part 4 looks at the debate from the perspective of two well-known ideas about how science works: by falsification (finding evidence inconsistent with the predictions from a proposed hypothesis) and paradigm shifts (sudden changes by the scientific community at large from one long-held view to a new governing paradigm). This section concludes that the community may now be in the early phase of a paradigm shift in how it thinks about past interactions between humans and climate, including the timing of forest clearance, greenhouse-gas increases, and effects on global temperature.

Part 5 assesses the corollary claim in the early anthropogenic hypothesis that Earth would by now have entered the first stage of a "glacial" world, with small ice sheets growing in several regions at high northern latitudes, had it not been for the effect of early agriculture keeping Earth warmer.

Part 6 investigates a second corollary claim that several small CO_2 and CH_4 decreases during the 2,000 years of the historical era were linked to episodes of mass human mortality caused by pandemics and civil strife. This section also assesses the possible contribution from these human calamities to the "Little Ice Age" cooling between the years 1200 and 1700.

The far-ranging debate that lies at the heart of this book provides an unusually clear example of how the process of science works. From eighty-some invited lectures at colleges, universities, and government labs during the last decade, I have found that students and the public at large are fascinated by real-world examples of science in action, particularly readily understandable examples like this one. This debate is also interesting because it draws on evidence from so many different disciplines, including archeology and archeobotany, ice-core geochemistry, marine geology, and the study of past climates. This information is introduced incrementally throughout the book as required to develop a basic line of argument.

Teaching from This Book

This book is written primarily for undergraduates at colleges and universities. It is suitable for a highly interdisciplinary course taught by instructors interested in providing a basic overview of the record of early human effects on landscapes and climate, with supplementary reading on topics of particular interest. It should be of interest to students majoring in disciplines such as earth and environmental science, archeology, ecology, and geography.

Another of my intentions in writing this book was for it to be used for students interested in humanities and social science courses but wishing to satisfy a science requirement or just pursue their own curiosity about human prehistory and history. Although this book covers many topics that will inform students about the operation of the climate system, it builds from simplicity toward moderate complexity in a step-by-step way. Any student majoring in English or history should be able to learn this material.

Instructors who choose to use *Earth Transformed* as a textbook have several choices. Because the book is relatively short, it can easily be covered in a semester-length course, with extra time available to dig more deeply into issues of greatest interest to the instructor. Additional resources listed at the end of each chapter can help inform choices of outside reading.

A second option is to teach only Parts 1–4, which cover the major issues relevant to the explanations of the greenhouse-gas trends during the last several thousand years. The summary at the end of Chapter 14 works as a natural endpoint to a course taught in this way.

A third option is to cover Parts 1–5, with Part 5 bringing in the question of whether small ice sheets would now be growing in several high-latitude regions in the Northern Hemisphere had farming not kept Earth warm by emitting greenhouse gases. The summary at the end of Chapter 17 works as a natural endpoint for such a course.

The chapters and sections vary widely in subject matter and degree of difficulty, although none rates as highly difficult:

- Part 1 is moderately complex because it introduces the wide range of background climatic and environmental information needed to set the stage for the mystery of the wrong-way greenhouse-gas trends.
- Part 2 is a largely descriptive treatment of the origin and spread of crops and livestock across the continents where early agriculture developed.
- Part 3 is the most complex section, because it digs deepest into the evidence on the two sides of the debate. But by this time, attentive readers will have learned enough to handle the depth of treatment.
- Part 4 is largely a description of how science works, as applied to this particular debate.
- Part 5 is a moderately complex treatment of whether or not the "next" glaciation is actually overdue. Again, readers will by this time be well prepared for the level of treatment.
- Part 6 is a moderately complex treatment of the claim that human calamities evident in historical records drove greenhouse-gas concentrations to lower levels.

For instructors and students looking for more information on past climates and the ways in which they are measured, my textbook *Earth's Climate: Past and Future* is available. In addition, key terms set off in bold in the text are explained in a glossary near the end of the book. I have avoided the jargon typical of much scientific writing. Jargon is often useful for shorthand communication among scientists, but it generally hinders understanding by nonexperts.

For example, earth scientists often refer to long stretches of time by using terms not familiar to general readers and students, such as: years BP (years before present) and kaBP ("kilo-annums" or thousands of years before present). Here, I use "years ago" to refer to these longer spans of time. Records from the historical era (the last 2,000 years) are labeled here as "year," where actual years are known, or by century. I avoided dividing time into BC (Before Christ) or the more recent BCE (Before Common Era), as well as AD (Anno Domini) or CE (Common Era). All of these formulations use a time around 2,000 years ago as a dividing line between what preceded and what followed; this usage causes confusion, such as having to mentally convert 1,000 BCE to roughly 3,000 years ago. Because this book focuses on the last 10,000 years, my general use of "years ago" keeps the flow of time intact. The use of "year" for showing some records from the historical era has the advantage of already being familiar to readers. Temperatures are generally listed in the normal scientific notation of degrees Celsius (°C) but also, in places, in degrees Fahrenheit (°F).

Finally, I use the same conventions I used in *Earth's Climate* for plotting trends against time. For very long records (tens to hundreds of thousands of years in length), I show the youngest at the top and the oldest at the bottom, the same sense in which layered records are deposited in sediments and ice. For shorter records (less than about 20,000 years), time is shown older on the left and younger toward the right, as in conventional historical records. All of these conventions seem to have worked well in *Earth's Climate*—I haven't heard any complaints!

ABOUT THE AUTHOR

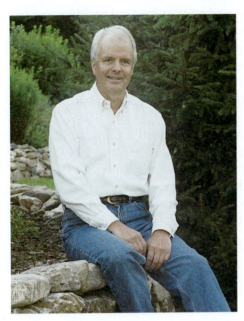

William F. Ruddiman holds a 1964 undergraduate degree in geology from Williams College and a 1969 Ph.D. in marine geology from Columbia University. He worked at the U.S. Naval Oceanographic Office from 1969 until 1976 as a senior scientist/oceanographer and then returned to Columbia's Lamont-Doherty Observatory as a senior research associate. He was associate director of the Oceans and Climate Division from 1982 to 1986 and adjunct professor in the Department of Geology from 1982 to 1991. He moved to the University of Virginia in 1991 as a professor in the Department of Environmental Sciences and served as department chair from 1993 to 1996. At Virginia he taught courses in climate change, physical geology, and marine geology. He has participated in fifteen oceanographic cruises and was co-chief on leg 94 of the Deep-Sea Drilling Project and leg 108 of the Ocean Drilling Project.

Professor Ruddiman's research interests have ranged across aspects of climate change on several time scales. His early research with Andrew McIntyre focused on orbital-scale climate change in the north and equatorial Atlantic Ocean. He was a member of the CLIMAP project from 1978 to 1984 and its project director from 1980 to 1981. He was a member of the COHMAP project from 1980 to 1989 and served on the steering committee throughout that time. Along with other COHMAP scientists, he coedited the 1993 volume *Global Climate Since the Last Glacial Maximum*.

In the late 1980s and the 1990s, his research focused mainly on the longer-term (tectonic-scale) physical and geochemical effects of uplift of the Tibetan Plateau and other high topography on regional and global climate. In 1997, he edited a book titled *Tectonic Uplift and Climate Change*, published by Plenum Press. His work on plateau uplift with colleagues Maureen Raymo, John Kutzbach, and Warren Prell has been featured in BBC and NOVA television documentaries.

Since 2001, his research has focused on the effects of early agriculture on greenhouse-gas concentrations. His early anthropogenic hypothesis proposes that releases of carbon dioxide and methane caused by farming drove an anomalous rise in greenhouse-gas concentrations during the last several thousand years. In 2005, his Princeton University Press book *Plows, Plagues, and Petroleum* won the Phi Beta Kappa award for best science book of the year.

Professor Ruddiman is also author of *Earth's Climate: Past and Future*, published in 2001, with new editions in 2007 and 2013. He is a fellow of the Geological Society of America and the American Geophysical Union. In 2010 he was awarded the Charles Lyell Medal from the Geological Society of London and in 2012 received the Distinguished Career Award from the American Quaternary Association.

DID CIVILIZATION DEVELOP IN A NATURALLY WARM WORLD?

Earth has existed in a glacial era for the last 35 million years. For much of the last 4.5 billion years, continent-sized **ice sheets** were not present anywhere on the planet. But by 35 million years ago, if not earlier, ice had appeared near the South Pole, and, for the last 14 million years, Antarctica has been almost completely entombed in ice.

The continents surrounding the Arctic Ocean are less frigid than Antarctica, but cooling had also been underway in the north for those same tens of millions of years, with massive ice sheets forming for the first time in northern Canada and Scandinavia by 2.75 million years ago. For most of the next 2 million years, those ice sheets grew and melted in regular 41,000-year cycles driven by changes in Earth's orbit around the Sun. Ice sheets covered parts of the northern areas of those continents during about half of those cycles and were absent during the other half.

Beginning nearly 900,000 years ago, the Northern Hemisphere moved deeper into a glacial world. In cycles averaging 100,000 years in length, ice sheets grew to much larger sizes and persisted on the continents for 90% of the time. The intervening ice-free intervals (**interglaciations**) lasted, at most, 10,000 years before the next long glacial phase returned.

Currently, thanks in part to those orbital changes, we live in one of those brief, warm, ice-free interglacial intervals that have been typical of only 10% of Earth's "average" **climate** for almost a million years. The most recent ice sheet on Scandinavia had disappeared by 11,000 years ago, and the last remnants of the much larger Canadian ice sheet melted away about 7,000 years ago. Although these changes may seem ancient from our day-to-day perspective, they overlap a vital part of the early history of human civilization. Not long before 10,000 years ago, as ice sheets were melting and the climate was warming, people in the warmer latitudes of Eurasia had begun the process of adapting wild plants and animals for farming. By the time the last ice melted in Canada, agriculture was spreading across large parts of the continents. The agricultural roots of our modern civilizations lie in the last stages of the most recent ice age.

Well after the last ice on North America and Europe had melted and the world had entered a time of interglacial warmth, our ancestors abandoned tools made of stone and instead began to manufacture bronze implements around 4,500 years ago (Figure P-1). This **Bronze Age** saw the appearance of early villages, towns, and even some urban centers, as well as the first known written records. By 2,500 years ago, tools made of bronze had begun to give way to iron implements in some regions. The subsequent **Iron Age** saw the founding of large premodern cities and major religions, as well as many technological innovations long before the explosive **Industrial Era** changes of the last two centuries. Many scientists have inferred that the warmth and stability of our current warm interglacial climate has been beneficial to the development of civilizations, in contrast to the colder, windier, dustier, and far more variable conditions that prevailed during the most recent glaciation.

At times during the current interglacial interval, climate changes in some regions dealt harsh setbacks to emerging cultures. Drought helped to decimate both the Akkadian cultures of the Tigris-Euphrates river valley 4,200 years ago and the Mayan cultures of Central America between the

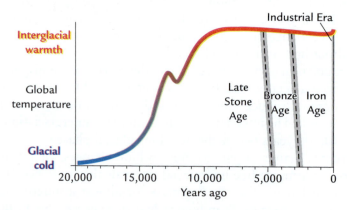

FIGURE P-1 The major eras of human prehistory as defined by tool use.

years 900 and 1000. Colder temperatures helped to eliminate Nordic peoples from Greenland after the year 1400. Yet these setbacks were brief in duration and local in extent, just short-term interruptions to ongoing human progress in a generally favorable climate.

The focus of this book will not, however, be the effects of climate on early human cultures, but rather the opposite: the possibility that humans have been altering Earth's climate for thousands of years through the unintended consequences of agriculture. This proposed early human overprint long predates the human intervention in the climate system during the current Industrial Era.

Decades ago, as an undergraduate student already interested in past climatic change, I was taught that the warmth of the current interglaciation was natural. Later, in graduate school, I learned that near-Arctic latitudes had cooled for several thousand years prior to the modern Industrial Era, but that the forces that drive long-term glacial cycles had not yet pushed the system across the threshold needed to initiate new ice sheets in Canada or Scandinavia. This view—that the warmth of the current interglacial is natural in origin—had prevailed during all the half century or so since studies of past climate first emerged as a distinct field.

Then, about a decade ago, I noticed something that didn't fit this view. The subsequent process of discovery was a bit like the old metaphor of pulling on a stray loose thread until a sweater completely unravels. But in this particular case, instead of piling up as useless yarn on the floor, the unraveled threads gradually rearranged themselves into a completely different garment—a new **hypothesis.**

In 2003, I wrote a paper challenging the conventional view that civilization emerged during the natural warmth of the last 10,000 years. Instead, I proposed that our farming ancestors began to interfere in the natural operation of the climate system long ago by adding **carbon dioxide** (CO_2) and **methane** (CH_4) to the atmosphere (Figure P-2A). I claimed that emissions of these **greenhouse gases** from early farming kept the climate relatively warm when natural trends would have caused a cooling. As millennia passed, and as agricultural practices spread across the planet, their warming influence on climate steadily grew, countering much of the slow natural drift toward cooler conditions. These warm interglacial millennia also saw the emergence of civilizations with urban areas, writing, and the world's major religions, but, according to the hypothesis, all of this occurred in a climate altered by human activities long before the Industrial Era.

Bold claims like this one need to be underpinned by solid evidence. As this book will explore, some of the evidence comes from knowledge about past climate changes and their causes. A key source of information

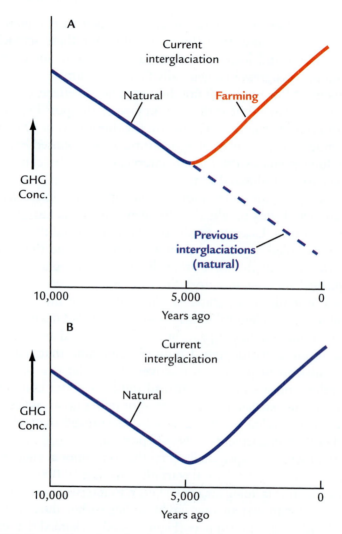

FIGURE P-2 Two explanations for the origin of the rise of greenhouse-gas concentrations during the last several thousand years: anthropogenic (A) and natural (B).

comes from astronomers who have calculated changes in the amount of radiation that each part of Earth's surface receives from the Sun during long-term changes in Earth's orbit. A second source comes from the natural records of climate information contained in ancient bubbles of air trapped in ice cores. These ice-core records tell us about changes in greenhouse gases during past climatic cycles driven by the changes in Earth's orbit.

Such sources provide crucial evidence of how Earth's climate behaved during times in the past, long before people could have influenced the climate system in a major way. As a result, the changes that occurred during these earlier times define natural climatic responses that can be compared against the trends during the present interglaciation. In effect, nature has run a series of gigantic natural experiments on our behalf that tell us what is natural behavior (and what is not). This evidence suggests that the current interglaciation has not followed the pattern expected from the previous ones that were driven by natural processes.

Still more evidence comes from countless archeological and historical records that document the spread of farming across Earth's surface. In the last several thousand years, agriculture has transformed the surfaces of the ice-free continents from a "natural" state to one largely altered by humans. Forests have been cut and replaced by croplands and pastures, and lowlands have been flooded to create irrigated rice paddies. These **anthropogenic** transformations of Earth's surface, which together constitute the largest changes humans have made to the surface of this planet, were constantly spreading during the same interval in which Earth's climate was staying warm rather than cooling. The new hypothesis views these early farming activities as the most likely explanation for Earth's "unnatural" climatic behavior.

As this book will show, this hypothesis has been vigorously criticized by a few scientists with first-rate reputations. They believe that the greenhouse-gas increases of the last few thousand years have been almost entirely natural in origin (see Figure P-2B). These two opposing explanations of the recent greenhouse-gas increases are still being debated.

But if this new hypothesis proves correct, it will change the context in which we view the history of civilization. In the previous view, humans have been the lucky but passive recipients of a favorably warm interglacial climate for many thousands of years. But in the new view, humans were at least partly responsible for maintaining the comfortably warm climate in which civilizations developed. Had it not been for the greenhouse-gas emissions from early agriculture, we would be living today in a cooler world. At some point during the last several thousand years, according to the hypothesis, small ice sheets would have begun forming in parts of Canada and northern Eurasia, and they would now be steadily growing and expanding.

Part 1 of this book describes how this new hypothesis emerged from initial indications that greenhouse-gas trends during past interglaciations differed from those during the present one (see Figure P-2A). Part 2 reviews archeological and historical evidence for the origin and spread of agricultural activity across the major continents during the last 10,000 years as a possible explanation for these "wrong-way" greenhouse-gas trends. These massive agricultural transformations of Earth's surface are one reason for the title of this book.

Part 3 focuses on the evidence behind the scientific disagreements over the relative role of natural versus anthropogenic explanations of the greenhouse-gas increases during the last several thousand years. Part 4 puts this debate into the broader framework of two concepts of how scientific progress occurs: by steady, careful hypothesis testing and falsification, or by sudden "paradigm shifts."

Part 5 explores one corollary prediction of the hypothesis—that growing agricultural greenhouse-gas emissions during the last several thousand years may have prevented the onset of a new glaciation. Part 6 explores a second corollary—the possibility that intervals of pandemics and civil strife that killed tens of millions of people during the historical era played a role in briefly reversing the warming effects from human activities and nudging Earth's climate back toward a slightly cooler state, the Little Ice Age.

The way science works is a mystery to most people, mainly because the background needed for understanding it is so complex. In contrast, the story told here is one that can be reasonably easily understood by those who have no prior acquaintance with the subjects covered, but who have open, curious minds. It is my hope that readers will gain a new understanding of both the origins of human civilization and of the scientific process from this book.

A Mystery: Wrong-Way Greenhouse-Gas Trends

For at least 90% of the past million years, ice sheets have covered parts of northern Canada, Scandinavia, and northern Asia. When the ice sheets reached maximum size, as they did most recently 20,000 years ago, they advanced far enough southward to cover the locations of modern-day Boston, New York, Chicago, Seattle, and Copenhagen. For much of the rest of each glacial interval, ice was present but less extensive, covering northeastern Canada, the mountains of Norway, and other regions in the far north. For the last million years, ice-free interglacial intervals like the present one have been short interruptions in a mostly glacial world.

These ice-age cycles are initiated by subtle changes in Earth's orbit around the Sun. These changes slowly alter the amount of **solar radiation** arriving at different latitudes during slow cycles lasting tens of thousands of years. The northern ice sheets then respond to these changes in high-latitude summer radiation. When the summer Sun is stronger, snow and ice melt; when it weakens, snow and ice accumulate.

The second major influence on the size of northern ice sheets has been changes in the concentration of two greenhouse gases in the atmosphere. Carbon dioxide (CO_2) and methane (CH_4) trap the radiation emitted by Earth's Sun-warmed surface, keeping it in the atmosphere, and making Earth's surface even warmer. Past concentrations of these gases can be measured by analyzing tiny air bubbles preserved in long ice cores extracted from the Antarctica and Greenland ice sheets. During the glacial cycles, both gases have generally acted to amplify the effects of changes in solar radiation. When the incoming summer sunlight strengthened and started to melt the ice sheets, atmospheric greenhouse-gas concentrations increased, and the added warmth the gases trapped in the atmosphere helped to melt even more ice. When the summer sunlight weakened, allowing ice sheets to grow, the gas concentrations in the atmosphere dropped, and the resulting cooler temperatures aided the accumulation of additional ice.

The current interglacial interval started off much like the previous ones. The large ice sheets that had developed by 20,000 years ago began to melt around 17,000 years ago as the summer sunlight slowly strengthened at high northern latitudes and as greenhouse-gas concentrations rose. By 10,000 years ago, the amount of summer radiation from the Sun had reached a peak roughly 8% stronger than the current value, and the northern ice sheets had largely melted. All the ice in northern Europe was gone by that time, and the remnant ice in Canada was shrinking rapidly toward a final disappearance about 7,000 years ago. Only the small ice sheet farther north on Greenland (and the massive one covering Antarctica) survived this warming.

But even before the last of the northern ice completely disappeared, the forces that had started ice melting had already reversed direction, pushing the climate system back toward cooler temperatures. After the peak in solar radiation 10,000 years ago, the strength of the incoming summer Sun at high northern latitudes began a long, slow decline that has continued up to the present day. This trend repeats an earlier pattern: similar decreases in summer solar heating also occurred early in previous interglaciations (Figure 1i-1A).

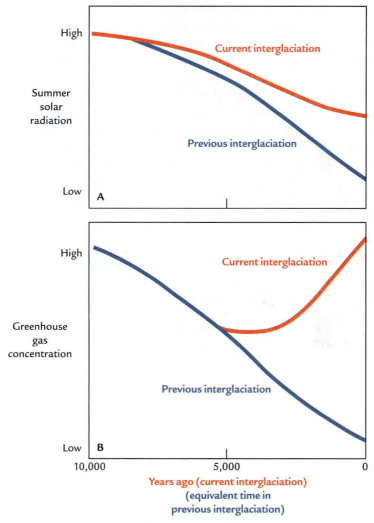

FIGURE 1i-1 The present interglaciation, as well as several previous ones, show similar downward trends in Northern Hemisphere summer solar radiation (A), but different trends in greenhouse-gas concentrations (B).

Concentrations of two greenhouse gases, CO_2 and CH_4, began by responding much as they had during previous interglaciations, reaching their peak nearly 10,000 years ago and then beginning to fall (Figure 1i-1B). Up to this point, these greenhouse-gas components of the climate system seemed to be behaving in a normal ("natural") way, tracking the decreasing trend in summer solar radiation and slowly moving Earth's climate toward the early stages of a cool glacial world.

But several thousand years ago, with summer solar heating continuing to decrease, the gas concentrations reversed direction and began to rise, veering off on a trend entirely different from the continued downward trends that had occurred in previous interglaciations. The CO_2 concentration reversed direction and began its anomalous rise close to 7,000 years ago, and the CH_4 concentration did the same close to 5,000 years ago.

It is difficult to explain these unexpected reversals and upward trends by calling on natural factors in the climate system. After all, if the natural gas trend is downward (as shown by the falling trends during previous interglaciations), how can the upward gas trends during the last few thousand years also be natural? This mismatch suggests that something new had begun to play a role in the climate system, a factor that was now intervening to produce this unexpected rise in these two greenhouse gases.

Additional Resources

Ruddiman, W. F. "The Atmospheric Greenhouse Era Began Thousands of Years Ago." *Climatic Change* 61 (2003): 261–293.

Ruddiman, W. F. *Plows, Plagues and Petroleum.* (Princeton: Princeton University Press, 2005).

Ruddiman, W. F. "How Did Humans First Alter Global Climate?" *Scientific American* (March 2005): 46–53.

CHAPTER 1

NATURE'S CLIMATIC CYCLES

Ducing the last few decades, investigations of past climatic change
have undergone a remarkable transformation. As recently as the
1950s, the coherent field of study called **paleoclimatology** barely
even existed. Individual scientists studied this or that aspect of climate
change, but were often disconnected from, or unaware of, other researchers in other specialties.

Contrast this situation with recent years, when a far-flung but
tightly interlinked international research community has come into existence: thousands of scientists study past and present climate at hundreds of academic and governmental institutions, using an array of
observing platforms (such as ships, satellites, and small unmanned
ocean floats) and ever more sophisticated lab instruments and comput-
ers. Interdisciplinary research has now become the norm, as scientists
have realized that the many parts of the climate system are intercon-
nected and need to be studied together.

In response to concerns about future global warming caused by
rising greenhouse-gas emissions, climate-related issues have even penetrated public consciousness in Hollywood films, such as the over-
wrought *The Day After Tomorrow*, starring Dennis Quaid as a
paleoclimatologist, and the documentary *An Inconvenient Truth*, fea-
turing former Vice President Al Gore.

The explosion in knowledge since the 1950s has been profound.
We now know that climate changes over all time scales, and we know
that these changes result from many kinds of **forcing** (drivers of change).
The climatic **responses** that result reflect the wide range of kinds of

forcing, from extremely slow shifts over tens to hundreds of millions of years (related to the gradual movement of tectonic plates and uplifting of plateaus and mountains) to changes that occur within a year or less (such as cooling episodes caused by volcanic explosions, and warm years produced by El Niño episodes in the Pacific Ocean). Much has also been learned about the balance of forces that drive all these changes, although many important questions remain unanswered. This scientific field is still relatively young.

Changes in Earth's Orbit

The climatic changes most relevant to the story explored in this book are caused by changes in Earth's orbit. Astronomers have calculated how the combined gravitational tugs of the Sun, the planets, and the moons of the planets affect Earth's path through space. Many decades ago, this work required laborious hand calculations, as there are seven planets in addition to Earth and dozens of smaller moons all moving through space, tugging on each other and on Earth. Powerful super-computers have sped up the process in recent decades to the point that reasonably accurate estimates of changes in Earth's orbit can now be made for millions of years into the past.

Two features of Earth's present-day orbit in space familiar to every-one are its once-a-year **revolution** around the Sun, and its once-a-day spin (**rotation**) on its axis. These movements are similar to those of a spinning top, which slowly traces out a more or less circular path (a revolution), and also spins many times during each complete revolution.

These present-day motions are the basis for understanding orbital changes that have occurred in the distant past. As Earth revolves around the Sun once a year, the axis around which it spins tilts at an angle of 23.5°, like the lean of a top (Figure 1-1). But Earth's angle of **tilt** has not always remained at 23.5°. Instead, it has ranged from a maximum of 24.5° to a minimum near 22°, changing from maximum to minimum and back to maximum at the rate of one full cycle every 41,000 years.

The shape of Earth's orbital path around the Sun has also changed. At present, the orbit is nearly circular in shape, but in the past the orbital path has usually been slightly more elliptical, with Earth located closer to the Sun at one point in its orbit than in the others (see Figure 1-1). A change between a more circular orbit and a more elliptical one, and then back toward a circular shape, referred to as a change in orbital **eccentricity**, occurs on average every 100,000 years.

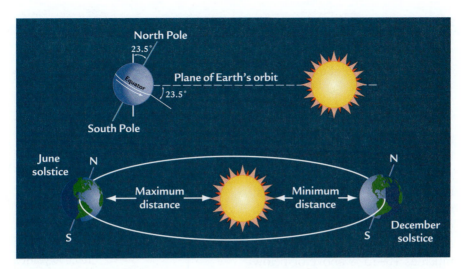

FIGURE 1-1 (top) Earth's rotational spin axis is now tilted at a 23.5° angle from a line perpendicular to the plane of its orbit around the Sun. (bottom) Because the eccentricity (shape) of Earth's orbit varies, changes in Earth's precessional motion (its "wobble") cause the seasons to occur at varying distances from the Sun.

Spinning tops also move in another way. By leaning in various directions at different times, they trace out a circular motion called a wobble. This wobbling motion in Earth's orbit is called **precession**. As Earth slowly wobbles through space, the direction in which it leans completes a full circle every 22,000 years. Note that while changes in orbital tilt through time refer to the *amount* of lean, changes in orbital precession refer to the *direction* of the lean.

Because of the interaction between the wobbling motion of precession and the elliptical orbital path, Earth is always closer to the Sun in a particular season and farther away in the opposite season. In the example shown at the bottom of Figure 1-1, Earth is closest to the Sun in December and farthest away in June. But these seasonal positions vary through time because of the wobbling motion. At times, Earth is closest to the Sun in June and farthest away in December.

These three changes in Earth's orbit are occurring constantly, but at different rates: 22,000 years to complete each full precession cycle, 41,000 years for each tilt cycle, and 100,000 years for each change in eccentricity. These same three cycles in turn affect the amount of solar radiation arriving on Earth, known as **insolation**. Insolation is measured in watts per square meter (W/m^2), the same unit used to report the amount of energy consumption in light bulbs (usually shortened to "watts").

FIGURE 1-2 Solar insolation received in June at 60°N over the last 300,000 years varied at cycles of 22,000 years and about 100,000 years.

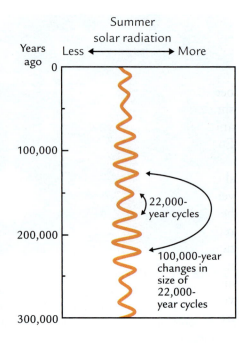

Note the orbital cycles present in Figure 1-2, which shows changes in northern hemisphere summer insolation at latitude 60°N during the last 300,000 years. The total range of insolation change through time at one place and for one particular season can amount to as much as 10–15% above or below the average value, equivalent to the present-day difference in solar radiation between Toronto and Atlanta, or between Oslo and Rome.

The total amount of insolation received by planet Earth doesn't change by much, because deficits in one hemisphere (northern or southern) are generally balanced by surpluses in the opposite one. Also, even though solar radiation is always much higher in summer than it is in winter, the extra amounts of insolation arriving in one season (whether summer or winter) are balanced by comparable deficits in the opposite season. During the early- to mid-twentieth century, most scientists felt that these opposing seasonal effects canceled each other, but later evidence has shown that this is not the case.

Instead, scientists now recognize that insolation changes in particular key seasons are more important than opposing changes in other seasons in driving orbital-scale changes in climate. As explained below, summer insolation in the Northern Hemisphere is particularly crucial in driving changes in ice sheets and summer monsoons, two very important components of the climate system.

Changes in Ice Sheets

Ice sheets are enormous bodies of ice several kilometers thick that cover major parts of continents. The two that exist today (on Antarctica and Greenland) have been relatively stable in size even during intervals of stronger solar heating, because they are located in very cold polar regions. In contrast, ice sheets that have existed in more temperate North America and Eurasia, farther from the North Pole, have been more vulnerable to cycles of changing solar insolation, repeatedly growing to large size and then melting away entirely. Canada has at times been covered by an immense mass of ice comparable in size to the ice sheet now covering Antarctica, while a smaller ice sheet has formed and then disappeared over the mountains of Scandinavia in northern Europe.

Decades of research have shown that these northern hemisphere ice sheets have varied in size at the three orbital cycles: 100,000 years, 41,000 years, and 22,000 years. The most recent melting of these ice sheets can be measured by the piles of debris left behind on land as the ice sheets retreated to the north, but this method cannot be used for earlier glaciations. Each successive ice-sheet advance scrapes and bulldozes the deposits from previous glaciations and piles them up in a jumble of debris along the new ice margins. Because the most recent ice advances were very large in extent, evidence of earlier ice sheets has been mostly erased from the land.

Somewhat surprisingly, the best record of ice sheets comes from the oceans, where sediments are deposited in continuous sequences that record the full history of successive glaciations. In ocean areas offshore of continental areas where ice sheets have grown and melted, the sedimentary record holds **ice-rafted debris**, the remains of mineral grains and rock fragments eroded by the ice, carried to the ocean by icebergs, and dumped into the deep sea (Figure 1-3). Even at a glance, sediment cores from these areas show the imprint of glacial cycles. Gray layers filled with continental minerals deposited from melting icebergs alternate with chalky white layers made up of shells of small marine **plankton** that flourished in warmer waters during ice-free interglaciations.

Ocean sediments also hold another even more useful index of the amount of ice on the continents. Two kinds of **oxygen isotopes** occur in the H_2O of seawater: a lighter oxygen-16 (^{16}O) version, and a heavier oxygen-18 (^{18}O) version. The snow that falls over growing ice sheets, and is incorporated into them as ice, comes from water in the ocean; the lighter isotope evaporated from seawater, the ^{16}O version, makes the long trip to the ice sheets more easily than the heavier ^{18}O isotope. Because more of the ^{16}O isotope is stored in the ice sheets, the water in the ocean is left with relatively more ^{18}O.

These changes are recorded in the shells of sand-sized plankton that float in the near-surface ocean and in organisms that live on the seafloor. As these creatures extract oxygen from seawater to form their

A

B

C

FIGURE 1-3 (A) Volcanic debris from Iceland and (B) red-stained quartz grains from (C) sandstone rocks around the Atlantic margins. *[A and B courtesy of G. Bond, Lamont-Doherty Earth Observatory of Columbia University. C: Alan Majchrowicz/Alamy.]*

calcium carbonate ($CaCO_3$) shells, they record the higher level of ^{18}O left behind in the ocean. After these creatures die, their shells pile up on the seafloor in layers that hold a continuous record of past ice volume. This oxygen-isotope index complements the layered sequences of iceberg-deposited debris in recording the waxing and waning of the great northern ice sheets (along with a much smaller contribution from Antarctic ice).

In 1976, the paleoclimatologists Jim Hays, John Imbrie, and Nick Shackleton compared a record of ice volume measured by oxygen isotopes in ocean sediments against independent astronomical calculations of Earth's orbit during the last 300,000 years. They found evidence that the northern ice sheets have responded to all three orbital cycles, and they concluded that the northern hemisphere summer is the critical insolation season controlling the size of the ice sheets, as predicted many decades earlier in the **Milankovitch hypothesis** proposed by Serbian mathematician Milutin Milankovitch.

This finding may sound counterintuitive, because much of the snow that builds ice sheets falls in the cold winter season. But in the High Arctic, winters are always very cold, even during warm interglacial climates like the current one. In such places, snow covers the ground for as much as 10 months of the year, melting away only in July or August before forming again in the beginning of autumn (September). In order for ice sheets to begin to form, some of that annual snowfall has to persist through the brief late-summer melt period.

Although today's warm summers manage to melt the snow, slightly cooler summers in the past under weaker incoming sunlight would have allowed some snow to persist through the melt season and form a base for additional accumulation during subsequent cold seasons. As snow piles up year after year, it gradually crystallizes into ice. When the ice reaches a thickness of 50–100 meters (about 150–300 feet), it can flow under its own weight as a glacier. In time, these growing glacial ice sheets can expand over enormous areas.

The reverse side of this process—ice melting—occurs when summer insolation in the Northern Hemisphere is strong. Nearly 20,000 years ago, vast ice sheets covered most of Canada and Scandinavia, but solar radiation levels were beginning to rise and the ice was about to start melting (Figure 1-4). By the time summer insolation reached a maximum, 10,500 years ago, the ice sheet over Scandinavia had already melted and the remnants of the much larger ice sheet on Canada were shrinking toward their final disappearance close to 7,000 years ago. (Because the ice sheet over Canada was so massive, it took several thousand years longer to melt in response to the stronger summer Sun.)

The two ice sheets that still exist today, covering Antarctica and Greenland, also hold valuable records of other climatic responses. Long sequences of cores drilled through several kilometers of ice tell us how greenhouse-gas concentrations changed in the past. The ice layers contain highly compressed bubbles of air that were trapped in fresh snow and gradually coalesced into layers of ice (Figure 1-5). Ice-core chemists

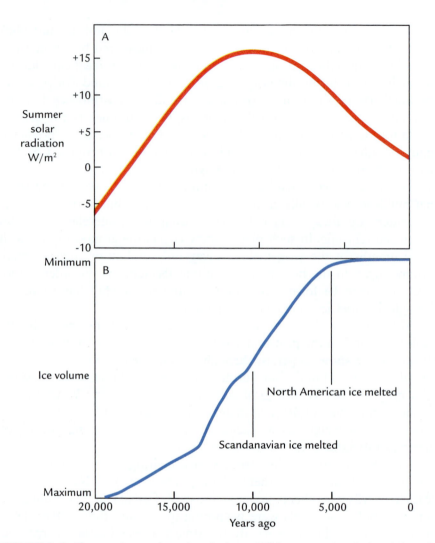

FIGURE 1-4 Changes in northern hemisphere 65°N summer insolation (A) and ice volume (B) during the last 20,000 years. *[Solar radiation trend adapted from A. Berger, "Long-term Variations of Caloric Insolation Resulting from the Earth's Orbital Elements,"* Quaternary Research *9 (1978): 139–167. Ice volume based on sea-level trend from E. Bard et al., "Calibration of the ^{14}C Time Scale Over the Past 30,000 Years Using Mass-spectrometric U-Th Ages from Barbados Corals,"* Nature *345 (1990): 405–410.]*

analyze gases like carbon dioxide (CO_2) and methane (CH_4) that are present in minute but measurable amounts in these bubbles of old air.

The concentrations of CO_2 and CH_4 vary with the cycles of Earth's orbit. Both gases tend to be more abundant when the climate is warmer and ice sheets are small, and less abundant when the climate is colder

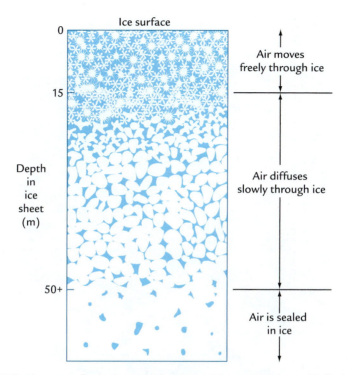

FIGURE 1-5 Air moves freely through the upper 15 meters (about 50 feet) of snow near the surface of ice sheets, but its flow is sealed off below a 50-meter depth. *[Adapted from D. Raynaud, "The Ice Core Record of the Atmospheric Composition: A Summary, Chiefly of CO_2, CH_4, and O_2," in* Trace Gases in the Biosphere, *ed. B. Moore and D. Schimmel (Boulder, CO: UCAR Office for Interdisciplinary Studies, 1992).]*

and ice sheets are large. Because of their **greenhouse effect** on the climate system, both gases tend to amplify climatic cycles started by changes in summer solar radiation. When summer insolation is strong and ice sheets melt, the greenhouse gases become more abundant and hasten the melting process. When summer insolation weakens and ice sheets start to form, the gas concentrations drop, helping the ice sheets grow.

Changes in Summer Monsoons

Another important part of Earth's climate system that responds to orbital changes is the circulation of the summer **monsoons**, which mainly operate in the warm tropics and subtropics far from regions of polar ice. At these latitudes, precipitation is a more important factor in the climate system than temperature, and the amount of precipitation that falls is mainly determined by the strength of the wet summer monsoons (Figure 1-6).

FIGURE 1-6 Drenching rains occur most summer afternoons in low-latitude monsoon regions. *[AFP/Getty Images.]*

Each summer, monsoon circulations are set into motion by the contrast in heating of the ocean and of the land. Strong heating of the land causes warm air to rise and pulls in moisture-bearing air from the nearby ocean (Figure 1-7). As the moist ocean air rises, it cools, and its water vapor reaches the **dew point** and condenses, falling in drenching afternoon monsoon rains. Changes in tropical circulation also occur in winter, but much less rain falls during the cooler, drier winter season.

Changes in past summer monsoon strength can be measured in several ways. The water levels of tropical lakes respond to changes in rainfall, so measurements of past changes in lake shorelines provide a good index of monsoon strength over the last several tens of thousands of years. Farther back in time, the oxygen-isotope index can be used, but in a different way from the example discussed previously. Changes in the $^{18}O/^{16}O$ isotopic composition of tropical rainfall are related to

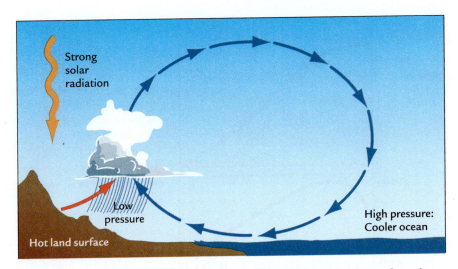

FIGURE 1-7 Strong solar heating of land surfaces in summer causes hot air to rise and draws in moist air from the nearby ocean, causing heavy rainfall.

monsoon intensity and are permanently recorded in limestone cave deposits made of calcium carbonate. These **stalactites** and **stalagmites** are formed by groundwater dripping down from Earth's surface as a result of drenching wet summer monsoons. As was the case for ice sheets at high latitudes, strong tropical monsoons deliver rainfall rich in ^{16}O from the ocean, and this abundance is recorded in the cave deposits.

In 1981, the atmospheric scientist John Kutzbach came up with a simple but elegant explanation of why these monsoon circulations varied in the past. He proposed that monsoon changes would have been an amplified version of seasonal changes observed today (Figure 1-8).

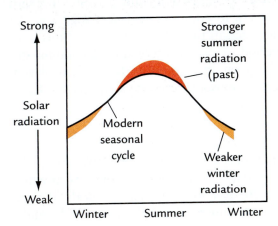

FIGURE 1-8 Past summer insolation levels higher than those today (red) drove monsoon circulation to be more intense than those today.

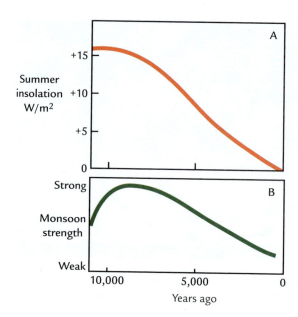

FIGURE 1-9 Summer insolation (A) and tropical monsoon strength (B) have been closely correlated for the last 10,000 years. *[Solar radiation trend adapted from A. Berger, "Long-term Variations of Caloric Insolation Resulting from the Earth's Orbital Elements,"* Quaternary Research 9 *(1978): 139–167. Tropical monsoon strength generalized from various sources.]*

Solar radiation is stronger in summer than in winter, and it drives the modern-day wet summer monsoons. At times in the past when summer solar radiation was higher than it is now, the land would have been heated even more in summer, driving a stronger monsoon inflow with greater rainfall than we see today. At times of weaker summer radiation, rainfall from the summer monsoons would have been weaker than today.

Recent monsoon trends support the **Kutzbach monsoon hypothesis.** Nearly 10,000 years ago, insolation levels in the Northern Hemisphere and monsoon strength in the northern tropics were both at peak values (Figure 1-9). Subsequently, summer insolation and monsoon strength fell slowly to their modern levels. In addition, evidence from cave deposits stretching much farther back in time shows that monsoons have varied at the 22,000-year period of orbital precession that dominates low latitudes. This match between the tempo of tropical insolation forcing and that of the tropical monsoon response supports the Kutzbach hypothesis.

Summary

Scientists have now arrived at a deeper understanding of two of Earth's great climate systems—ice sheets and monsoons—and the way in which both have varied in response to past orbital changes. Both of these systems are highly sensitive to summer solar radiation, which guides

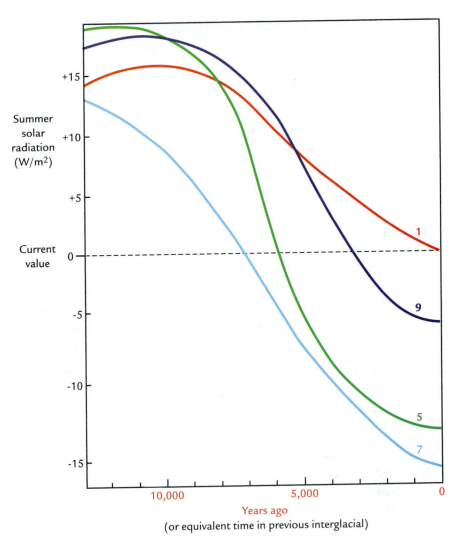

FIGURE 1-10 Changes in summer insolation early in the present interglaciation (Stage 1), compared to changes during three preceding interglaciations. *[Trends are from A. Berger, "Long-term Variations of Caloric Insolation Resulting from Earth's Orbital Elements,"* Quaternary Research 9 (1978): 139–167.]

the growth and melting of northern ice sheets, and the strengthening and weakening of tropical summer monsoons. Neither system is strongly sensitive to insolation changes in winter.

Of greatest importance to the story explored in this book are the summer insolation changes that occurred early in interglaciations, when

TABLE 1-1 **Ages of Interglaciations**

Name of interglaciation*	Age (years ago)
Stage 1	Present-day
Stage 5	120,000
Stage 7	240,000
Stage 9	330,000

*By convention, interglacial stages with warm climates have odd numbers and glacial stages with cold climates have even numbers.

the last remnants of the ice sheets were melting and warm interglacial conditions were taking hold. For the present interglaciation (named "Stage 1") and the three previous interglacial stages (ages shown in Table 1-1), the summer insolation trends all look similar: peak values at the start of each interglaciation, and then a slow drift toward lower values (Figure 1-10). The next two chapters address the question of whether CO_2 and CH_4, both important greenhouse gases, have also followed similar trends early in interglaciations.

Additional Resources

Berger, A. "Long-term Variations of Caloric Insolation Resulting from the Earth's Orbital Elements." *Quaternary Research* 9 (1978): 139–167.

COHMAP Project Members. "Climate Changes of the Last 18,000 Years: Observations and Model Simulations." *Science* 241 (1988): 1043–1052.

Ruddiman, W. F. *Earth's Climate: Past and Future.* (New York: W. H. Freeman, 2007), Chapters 7–9.

CHAPTER 2

WRONG-WAY METHANE TREND

Methane (CH$_4$) is an important greenhouse gas produced by both nature and human activities. Most natural methane originates in wetlands, where large amounts of carbon-rich vegetation grow in standing water during the warmth of summer (Figure 2-1). The majority of other environments on Earth are rich in oxygen, and carbon in dying vegetation is oxidized to carbon dioxide (CO$_2$), which enters the atmosphere. But in wetlands, plants made of carbohydrates (CH$_2$O) are attacked by bacteria and slowly decompose. The decay process consumes oxygen, leaving the water oxygen-free, and the carbon and hydrogen from the dead plants produce methane (Figure 2-2). Those bubbles of "swamp gas" or "marsh gas" you may have seen bubbling up out of stagnant pools of water are mostly methane.

Geochemists use bubbles of air trapped in ice cores to measure past methane concentrations in Earth's atmosphere in units of parts per billion (**ppb**). Measurements spanning the last 15,000 years show initially low values late in the most recent glaciation (Figure 2-3). Concentrations then rose rapidly until briefly interrupted by a return to lower values during a cold climatic event called the **Younger Dryas** around 12,000 years ago, followed by an abrupt rebound to peak values by 11,000 years ago. From 11,000 to 5,000 years ago, methane concentrations fell, but then they rose again through the present day. The CH$_4$ increase over the last several thousand years is the focus of this chapter.

FIGURE 2-1 Wetlands support abundant carbon-rich vegetation.
[*All Canada Photos/Alamy.*]

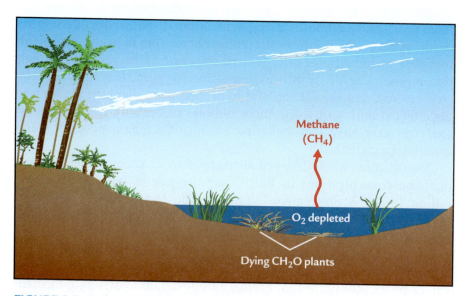

FIGURE 2-2 Carbon-rich vegetation growing in wetlands rots in oxygen-depleted water and releases CH_4 to the atmosphere.

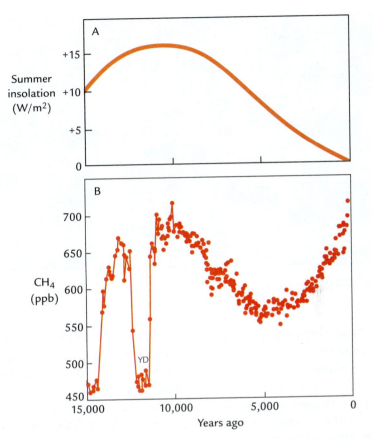

FIGURE 2-3 Trends in summer insolation and atmospheric methane concentrations during the last 15,000 years. *[Insolation adapted from A. Berger, "Long-term Variations of Caloric Insolation Resulting from Earth's Orbital Elements," Quaternary Research 9 (1978): 139–167. CH₄ values adapted from EPICA Community Members, "Eight Glacial Cycles from an Antarctic Ice Core." Nature 429 (2004): 623–628.]*

Methane from Monsoon Wetlands

One factor that explains the overall shape of the CH_4 trend (except for the Younger Dryas interruption) is the changing strength of northern hemisphere summer monsoons. Two factors enhance summer monsoon rainfall: large continental areas that can be strongly heated by summer insolation, and high topography that can serve as a focal point for especially heavy precipitation. Strong summer monsoon precipitation fills natural wetlands that emit methane.

Asia has the world's strongest summer monsoon circulation (Figure 2-4) because it is the largest continent on Earth and has very high topography

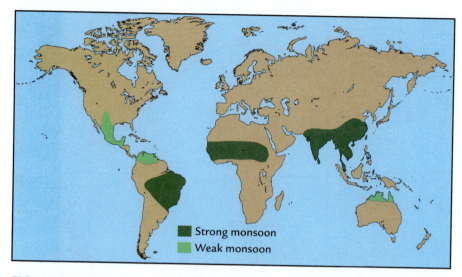

FIGURE 2-4 Large tropical and subtropical regions are influenced by wet summer monsoons that flood natural wetlands.

along its southern margin—the Himalaya Mountains and Tibetan Plateau. Strong summer heating of this giant land mass initiates the Asian monsoon circulation (see Chapter 1, Figure 1-7), which delivers particularly heavy rains to the south Asian mainland.

Africa is also a large continent, but the topography of the area north of the equator is not as high as southern Asia, so the North African summer monsoon rains are somewhat less intense and extensive than those in Asia. The Amazon Basin of South America is also an important summer monsoon region (see Figure 2-4), but it is smaller than the combined size of the northern monsoon regions, and its contribution to the global monsoon circulation is also smaller.

Chapter 1 showed how changes in summer insolation drive monsoon circulations, and now we can add another link to that cause-and-effect chain: summer monsoon rains create wetlands that emit methane. This link provides a plausible explanation for much of the methane record shown in Figure 2-3. Rising summer insolation levels after 15,000 years ago drove stronger summer monsoons that flooded methane-emitting wetlands and pushed CH_4 concentrations to a peak 11,000 to 10,000 years ago.

Observations from field studies in northern tropical regions confirm that this was a time of very strong monsoons (Figure 2-5). Old shorelines show that lake levels 10,000 years ago were much higher than they are now across the southern margin of the Sahara Desert. Deposits of sediment show that streams flowed across now hyperarid parts of the northeast

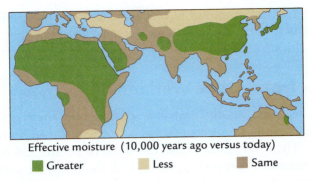

Effective moisture (10,000 years ago versus today)

■ Greater   ■ Less   ■ Same

FIGURE 2-5 Lake-level evidence indicates stronger monsoons 10,000 years ago across Africa and Asia compared to today. *[Adapted from COHMAP Members. "Climatic Changes of the Last 18,000 Years: Observations and Model Simulation," Science 241 (1988) 1043–1052.]*

Sahara (Sudan and Egypt) as well as southern Arabia, providing habitat for a rich array of wildlife that included hippopotami, crocodiles, and other animals dependent on reliable sources of water. Grass grew in intervening areas where desert sand blows today. The area of strong monsoons extended all the way east into southern China, where tropical forests adapted to heavy rainfall extended well north of their present-day limits.

After 10,000 years ago, northern hemisphere summer insolation began to decrease, monsoons weakened, and methane emissions fell until 5,000 years ago, along with the ice-core CH_4 trend (see Figure 2-3). Once again, this pattern is consistent with the proposed link between insolation, monsoons, and methane emissions. As the summer monsoon rains weakened, climates that had been seasonally moist gradually became more arid. In drier regions, grasslands gave way to desert sands, and lakes disappeared completely. In semi-arid regions, forests were gradually replaced by grasslands or **savanna** (a mix of grass and scattered trees), and lakes fell to lower levels. In wetter regions such as India and southern China, wet-adapted trees were replaced by dry-adapted ones. As the vast regions of southern Asia and North Africa slowly dried out, they must have emitted progressively smaller amounts of methane.

Until about 5,000 years ago, these insolation and CH_4 changes seem consistent with John Kutzbach's monsoon hypothesis. Summer insolation continued to decrease over the last 5,000 years, and field evidence from southern Asia and North Africa shows that the regional drying continued as expected with a weakening monsoon. But then something unexpected happened. The methane signal did not follow the anticipated downward trend; instead, it reversed direction and rose steadily right up through the start of the Industrial Era two centuries ago (see Figure 2-3). Northern-tropical wetlands were drying out and clearly could not have been the source of this CH_4 increase. But if they weren't the cause, what was?

Methane from Arctic Wetlands

Scientists considered another possible explanation for the increase of CH_4—wetlands in Arctic or near-Arctic (sometimes called *boreal*) regions (Figure 2-6). These wetlands lie too far north to be affected by tropical–subtropical summer monsoon rains; instead, the northern wetlands exist because evaporation is low during the cool short Arctic summers. Water left over from the melting of winter snow and moisture from occasional summer rains leaves low-lying areas across these landscapes dotted with ponds of standing water and water-saturated **peat bogs**. High-latitude summers warm up enough to allow vegetation to grow, and, once again, plant decomposition produces large amounts of methane.

At first, these regions seem to be plausible sources of the CH_4 increase after 5,000 years ago. Wetlands began spreading across Arctic regions nearly 13,000 years ago as the ice sheets melted back and the far north was released from frigid glacial conditions. These wetlands have continued to expand since 5,000 years ago, especially in Canada around Hudson Bay. Couldn't these expanding northern wetlands have

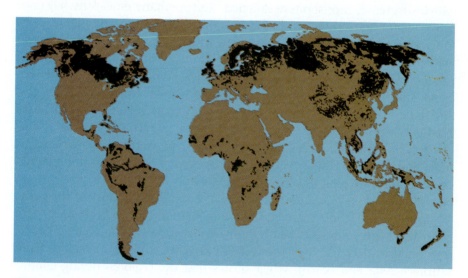

FIGURE 2-6 Large wetlands cover cold Arctic regions. *[From Z. Yu, J. Loisel, D. P. Brosseau, D. W. Beilman, and S. J. Hunt, "Global Peatland Dynamics Since the Last Glacial Maximum," Geophysical Research Letters 37 (2010), L13402, doi: 10.1029/2010GL043584.]*

generated increasing amounts of methane, thereby explaining the rise in atmospheric CH_4 concentrations?

North–South Differences in Methane Concentration

Evidence from ice cores tells us that the answer to this question is *no*. An Arctic methane source can be firmly ruled out based on differences in the concentration of CH_4 measured in ice drilled in Greenland versus ice drilled in Antarctica. Differences in concentrations in the two polar regions have always existed because of the unequal distribution of methane sources across the planet—more sources lie north of the equator than south of it because of the much greater area of northern continents compared to the more ocean-covered Southern Hemisphere (recall Figures 2-4 and 2-6).

This unequal geographic distribution of methane sources is important because of the relatively short lifetime of CH_4 in the atmosphere. Methane emitted from wetland sources only persists for an average of ten years or so before reacting chemically with oxygen and being transformed into other gases like CO (carbon monoxide) and eventually CO_2 (carbon dioxide). If methane stayed in the atmosphere for hundreds of years, its concentration would be evenly mixed across the planet, but this is not the case. Instead, its short lifetime allows regional differences in concentration to develop.

Because larger CH_4 sources exist in the Northern Hemisphere, and because it takes some time for northern air masses to move far south, some of the methane in the air is transformed into other gases before it reaches southern polar latitudes. As a result, atmospheric concentrations at the South Pole are a few percent lower than those over Greenland.

This north–south difference is largest for methane-laden air masses from Arctic sources. Methane emitted from Arctic sources into northern air masses has little time to decompose before it arrives over Greenland, leaving relatively high CH_4 concentrations trapped in Greenland ice air bubbles (Figure 2-7A). But parcels of air containing Arctic methane take much longer to reach south-polar latitudes, and more of the methane has been altered to other gases by the time the air is trapped in Antarctic ice. As a result, CH_4 concentrations in Antarctic ice are markedly lower than those in Greenland ice.

Methane emitted from the tropics and lower subtropics is a different case. Because low-latitude monsoon sources are nearly equidistant from Greenland and Antarctica, the air parcels heading poleward take similar lengths of time to reach the two ice sheets, and the CH_4 concentrations are reduced by similar amounts (see Figure 2-7B). As a result,

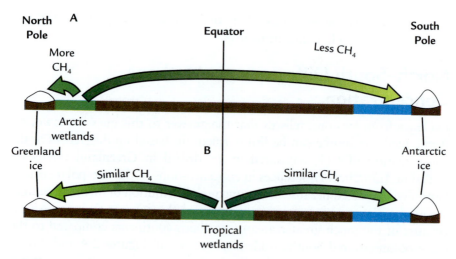

FIGURE 2-7 Methane emitted from wetlands in the Arctic (A) and tropics (B) travel different distances to reach the Greenland and Antarctic ice sheets and produce different concentrations.

methane emitted from the tropics and subtropics produces similar CH_4 concentrations in Greenland and Antarctic ice sheets.

These differences in methane sources and paths of poleward travel provide climate scientists with a way to figure out relative changes in the contributions from the two sources. Intervals when methane concentrations in Greenland ice grow large compared to those in Antarctic ice indicate that relatively greater CH_4 emissions are coming from Arctic sources close to Greenland. Conversely, intervals with reduced CH_4 differences between the two ice sheets indicate relatively greater emissions from the tropics and subtropics than from sources in the Arctic.

Over the last 5,000 years, as the overall methane concentration in the atmosphere has increased, the *difference* in CH_4 concentration in Greenland ice versus Antarctic ice has slowly decreased. This decreasing trend means that CH_4 emissions from wetlands in Siberia and Canada were falling, not rising. From this evidence, Arctic wetlands could not have been the source of the CH_4 increase.

At first, this conclusion doesn't seem to make sense. Why would emissions from the northern sources have fallen while the total area of wetlands expanded? The reason appears to be climatic. A wide range of evidence from the Arctic shows that climate was cooling during this interval, especially in the summer. Forests were retreating southward, episodes of summer melting recorded in small Arctic glaciers and ice caps were becoming less frequent, and sea ice was advancing southward. This cooling must have shortened the already brief summer growing season in the Arctic enough to suppress methane emissions from the vast (and still-expanding) peatlands in the far north. In any case, the methane differences between Greenland and Antarctic ice firmly eliminate Arctic wetlands as a major methane source that could explain the overall CH_4 increase 5,000 years ago.

Comparison to Methane Trends in Previous Interglaciations

When we consider methane trends during equivalent intervals in three preceding interglaciations, the mystery only deepens. The early parts of these previous interglaciations all show patterns similar to the current one: low CH_4 values during the preceding deglaciations, followed by increases to maximum value early in each interglaciation, and then a subsequent decrease (Figure 2-8).

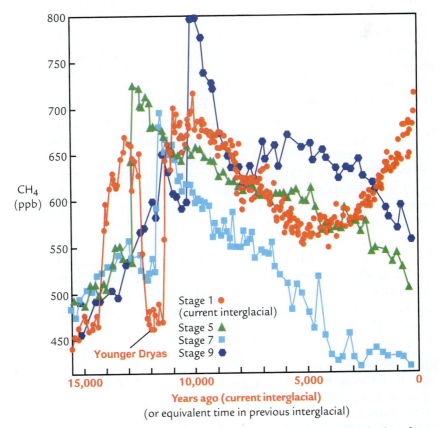

FIGURE 2-8 Trends in atmospheric CH_4 concentrations during the last four interglaciations. *[Adapted from EPICA Community Members, "Eight Glacial Cycles from an Antarctic Ice Core." Nature 429 (2004): 623–628.]*

But here the parallel behavior ends. CH_4 trends during the three previous interglaciations kept falling until the times equivalent to the present day, but the methane trend in the current interglaciation reversed direction and rose steadily during the last 5,000 years. The ongoing CH_4 decreases in the previous interglaciations were fully consistent with the continuing drop in summer insolation (see Chapter 1, Figure 1-10), but the CH_4 increase during the last 5,000 years is not.

Summary

These findings pose a major enigma. The increase in atmospheric methane that began 5,000 years ago is opposite in direction to the decreases that prevailed during the most similar times in the three previous interglaciations. And this "wrong-way" increase in methane did not come from either of the two natural CH_4 sources thought to be the largest on the planet (northern-tropical and Arctic wetlands). Both wetland sources were emitting progressively less methane during the last 5,000 years.

Even more baffling is the fact that methane concentration differences between Greenland and Antarctic ice indicate that the additional methane needed to explain the atmospheric increase during the last 5,000 years must have come from low (tropical–subtropical) latitudes rather than far-northern Arctic regions. Yet the immense natural wetland source regions in the tropics and subtropics of Eurasia and North Africa were drying out and emitting less methane during this interval, instead of becoming wetter and emitting more.

How can this mystery be explained? One possibility is that something new had appeared in the climate system during the last several thousand years, a new factor that had not previously been important but now began to override Nature's previous control of global methane emissions. One candidate for this mysterious methane trend is human activity: specifically, early agriculture. Most of our very early farming ancestors several thousand years ago lived in the lower latitudes of Eurasia, Africa, and the Americas. Could their early farming activities have produced the large amounts of tropical methane needed to answer this mystery?

Additional Resources

Blunier, T., J. Chappellaz, J. Schwander, B. Stauffer, and D. Raynaud. "Variations in Atmospheric Methane during the Holocene Epoch." *Nature* 374 (1995): 46–49.

Chappellaz, J., T. Blunier, S. Kints, A. Dallenbach, J.-M. Barnola, J. Schwander, D. Raynaud, and B. Stauffer. "Changes in the Atmospheric CH_4 Gradient

Between Greenland and Antarctica During the Holocene." *Journal of Geophysical Research* 102 D (1997): 15,987–15,997.

Kutzbach, J. E. "Monsoon Climate of the Early Holocene: Climate Experiment with Earth's Orbital Parameters for 9000 Years Ago." *Science* 214 (1981): 59–61.

Kutzbach, J. E. and F. A. Street-Perrott. "Milankovitch Forcing of Fluctuations in the Level of Tropical Lakes from 18 to 0 kyr BP." *Nature* 317 (1985): 130–134.

Ruddiman, W. F. *Earth's Climate: Past and Future*. New York: W. H. Freeman, 2007, Chapters 8, 13.

Ruddiman, W. F. and J. S. Thomson. "The Case for Human Causes of Increased Atmospheric Methane over the Last 5000 Years." *Quaternary Science Reviews* 20 (2001): 1769–1777.

CHAPTER 3

WRONG-WAY CARBON DIOXIDE TREND

Another greenhouse gas generated both by nature and by human activities, carbon dioxide (CO_2), plays a more important role in the climate system than methane for two reasons: it is much more abundant and it stays in the atmosphere much longer. Changes in CO_2 have a much smaller warming impact per molecule than methane, but the atmosphere contains nearly a thousand times as many molecules of CO_2 as methane. As a result, the net impact of CO_2 on past changes in climate has been considerably larger than that of methane.

Carbon Reservoirs

Billions of tons of carbon exist in several major carbon reservoirs that compose Earth's **biosphere** (Figure 3-1). Prior to the onset of Industrial-Era CO_2 emissions from fossil fuels, three surface reservoirs contained large amounts of carbon: 600 billion tons were held as **inorganic carbon** in CO_2 gas in the atmosphere, 610 billion tons of **organic carbon** were stored in above-ground vegetation, and 1,000 billion tons of mostly inorganic carbon were dissolved in well-mixed, sunlit upper layers of the ocean. Carbon moved back and forth among these three surface reservoirs relatively rapidly (over years to decades).

Two much larger reservoirs are relatively isolated from the climatic processes occurring at Earth's surface and exchange carbon much more slowly: roughly 1,550 billion tons of organic carbon are buried in continental soils, and some 38,000 billion tons of mostly inorganic carbon are dissolved in the deep ocean. Because of carbon emitted by human activities since the start of industrialization 150 to 200 years

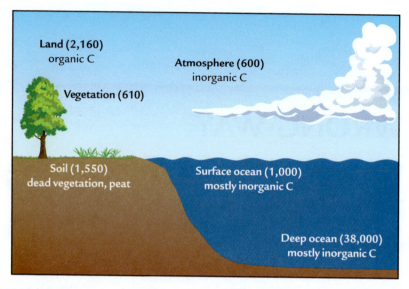

FIGURE 3-1 The major reservoirs of carbon on Earth (values shown in billions of tons, prior to carbon emissions during the Industrial Era).

ago, both the surface and deep ocean reservoirs now contain more carbon than the amounts shown in Figure 3-1.

Carbon moves back and forth between these various reservoirs primarily through two processes. Plants use energy from sunlight and nutrients from the environment to convert inorganic carbon (CO_2) to organic carbon (plant tissue) in a process called **photosynthesis**. The process of oxidation works in the opposite direction, converting organic carbon to inorganic carbon dioxide ($C + O_2 \rightarrow CO_2$).

Past concentrations of carbon dioxide can be measured in bubbles of ancient air trapped in ice cores, using units of parts per million (**ppm**). Over the last several hundred thousand years, CO_2 concentrations have varied similarly to changes in the size of ice sheets (Figure 3-2). CO_2 values were low when ice sheets were large, and high when ice sheets were small. Both the CO_2 and ice-sheet changes occurred in a "saw-toothed" pattern: the ice sheets grew slowly over many tens of thousands of years, but melted in less than 10,000 years. CO_2 concentrations moved in tandem with the ice sheets, gradually falling as ice sheets grew, and rising more quickly when they melted.

The reason for these similar changes in CO_2 and ice sheets is not yet entirely clear, but it seems to involve some kind of **feedback** relationship. The amount of ice has had an effect on CO_2 levels, and CO_2

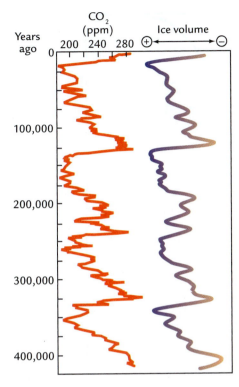

Years
ago

CO$_2$
(ppm)

Ice volume

FIGURE 3-2 Variations in
atmospheric CO$_2$ concentrations and
ice volume have followed similar
trends for hundreds of thousands of
years. [Adapted from J. R. Petit et al.,
"Climate and Atmospheric History of the Past
420,000 Years from the Vostok Ice Core,
Antarctica," Nature 399 (1999):
429–436.]

levels have had an effect back on the size of the ice sheets. Like the classic chicken-and-egg problem, it's hard to say which changed first.

The CO$_2$ trend for the last 15,000 years is generally similar to the methane trend, except that it does not show a major drop during the Younger Dryas. Starting with low values near 15,000 years ago (late in the most recent glaciation), the CO$_2$ concentration rose to a maximum near 11,000–10,500 years ago at the same time as the peak in methane (Figure 3-3). Then, like the methane trend, the CO$_2$ concentration slowly decreased for a few thousand years. But around 7,000 years ago, it reversed direction and rose until the present day, similar to the reversal in the methane trend that occurred 5,000 years ago (see Chapter 2, Figure 2-3).

Where Did the Missing Carbon Go During Glaciations?

One way to investigate these changes is to find out where all the extra CO$_2$ held in the atmosphere during interglacial intervals like today went during glacial periods. During times when ice sheets were at maximum size, like the period that occurred 20,000 years ago, the atmosphere only held about 420 billion tons of carbon (compared to 600 billion tons during interglaciations), so about 180 billion tons of carbon must have been transferred somewhere else. But where?

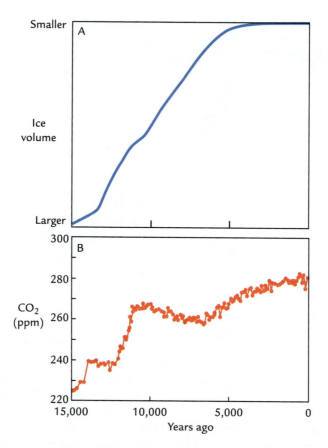

FIGURE 3-3 Changes in ice volume (A) and atmospheric CO_2 concentrations (B) during the last 15,000 years. *[Ice-volume trend based on sea-level rise from E. Bard et al., "Calibration of the [14]C Time Scale Over the Last 30,000 Years Using Mass Spectrometric U-Th Ages from Barbados Corals," Nature 345 (1990): 405–410. CO_2 values from EPICA Community Members, "Eight Glacial Cycles from an Antarctic Ice Core," Nature 429 (2004): 623–628.]*

We know that this carbon didn't go into the glacial-age forests, because the northern ice sheets covered vast areas of Canada and Scandinavia where forests exist today. Elsewhere, glacial-age forests were generally smaller in extent and the vegetation was less dense than today, because the climate was colder and drier, and because lower CO_2 values in the atmosphere reduced photosynthesis rates and the amount of vegetation. Also, lower vegetation cover naturally led to lower carbon storage in the underlying soils of most regions. The combined effect of reduced forest vegetation and reduced soil carbon adds roughly another 530 billion tons of carbon to the total that must have been stored somewhere else.

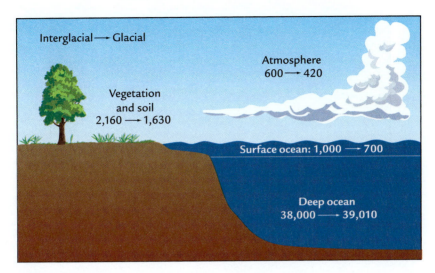

FIGURE 3-4 Changes in carbon stored in Earth's reservoirs from warm interglacial intervals to cold glacial intervals (in billions of tons).

We also know that the carbon didn't move into the surface ocean. Because the upper ocean layers exchange CO_2 with the overlying atmosphere over years to decades, the average amount of carbon in the global surface ocean does not stray far from the concentration in the atmosphere. Given this close link, the carbon concentration in the surface ocean during the last time when ice sheets were large must also have been lower than it was during interglaciations, by about 300 billion tons. Combined with the deficits in the atmosphere and land vegetation, the amount of "missing carbon" totals just over 1,000 billion tons (180 + 530 + 300 = 1,010).

So where on Earth did 1,000 billion tons of carbon go? The major reservoir left—the deep ocean—holds more than 60 times the amount of carbon in the atmosphere (Figure 3-4). Several kinds of evidence suggest that the extra carbon that had been held in the atmosphere, the vegetation, and the surface ocean during warm interglacial climates moved into the deep ocean as glaciations developed, and then returned to the surface reservoirs once climate warmed and ice sheets melted.

Scientists are exploring several mechanisms by which carbon could move into and out of the deep ocean during glacial cycles, and two ideas seem the most promising. One possibility is that floating algae (**phytoplankton**) that live in ocean surface waters were sending more carbon down to the deep ocean during glacial times (Figure 3-5A). Like

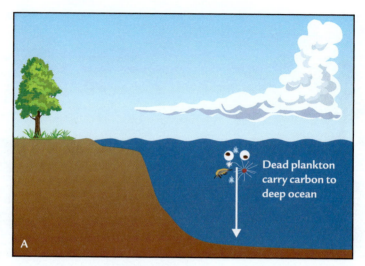

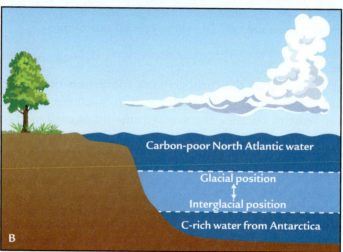

FIGURE 3-5 Carbon stored in the deep ocean can vary with changes in (A) the amount sent to the seafloor in the soft tissue of surface-water phytoplankton, and (B) circulation of deep-water masses.

all forms of life, these floating organisms carry organic carbon in the soft tissue of their bodies. When they die and their shells sink, some of this organic carbon is carried down into the deep ocean. The strength of this process can vary through glacial cycles, because changes in climate produce different wind patterns that drive different circulation patterns in the upper ocean layers, such as changes in **upwelling** of water from below. Also, windy glacial climates may have delivered more iron-rich dust from the land to the sea, which can act as a "vitamin supplement" that fertilizes surface waters and enhances algae growth, leading to increased amounts of sinking algae.

Another possible mechanism for varying the amount of carbon stored in the deep ocean is a change in deep-ocean circulation, especially in the Atlantic Ocean (Figure 3-5B). Deep water that forms in polar or near-polar regions carries carbon down into the deep ocean, and the rate of formation of deep water in polar regions varies in response to changes in climate (primarily the temperature and salinity of ocean water). As a result, the amount of carbon carried down can increase during colder glacial climates, as compared to warm interglacial climates. These circulation changes can also alter the amount of $CaCO_3$ (in the form of soft chalk) dissolved by acidic deep waters from sediments on the seafloor. All of these processes can result in more or less carbon being stored in deep ocean waters.

In any case, despite some gaps in scientific understanding, scientists know that low CO_2 values in the atmosphere occurred during glacial climates at least in part because extra carbon was stored in the deep ocean, and that subsequent increases in atmospheric CO_2 occurred during interglaciations because carbon was being released from the deep ocean back into the atmosphere as glaciations ended. The atmospheric CO_2 increase and ice-volume decrease that occurred from 15,000 to 11,000 years ago is consistent with this long-term pattern of carbon emerging from deep-ocean storage after the large glaciations ended (see Figure 3-3).

Unexpected CO_2 Changes During the Last 7,000 Years

The small CO_2 decrease that occurred between 11,000 and 7,000 years ago (recall Figure 3-3A) has been attributed at least in part to the last stages of melting of the northern ice sheets (Figure 3-6). As forests in Canada and Scandinavia shifted northward and colonized the land exposed by the melting ice sheets, atmospheric carbon was stored in the woody tissue of the trees. In addition, large amounts of carbon were being deposited in peatlands that were expanding across high-Arctic latitudes at that time (see Chapter 2, Figure 2-6).

But the 22-ppm rise in CO_2 that began near 7,000 years ago is a mystery (see Figure 3-3). This increase represents a massive transfer of carbon. Even though CO_2 floats in the atmosphere as a gas, it is not weightless. Each 1 ppm of atmospheric CO_2 concentration represents

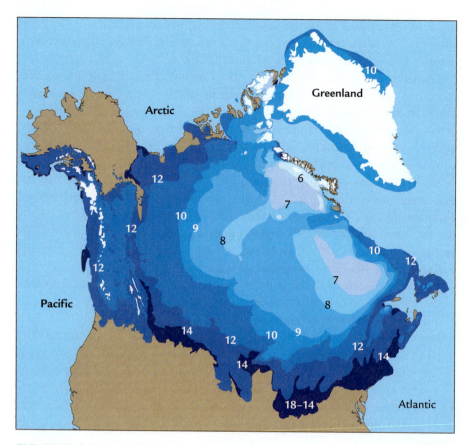

FIGURE 3-6 Dating of organic remains shows the margins of the ice sheets in North America retreating after 18,000 years ago, and melting away entirely near 7,000 years ago. Numbers on the map are in thousands of radiocarbon years, which are somewhat younger than actual calendar years. [*Adapted from A. S. Dyke and V. K. Prest, "Late Wisconsinan and Holocene History of the Laurentide Ice Sheet,"* Geographie Physique et Quaternaire *41 (1987): 237–263.*]

about 2.13 billion tons of CO_2, so the 22-ppm increase in CO_2 during the last 7,000 years requires an addition of almost 47 billion tons of CO_2 to the atmosphere (22 ppm $\times$ 2.13 tons/ppm).

This transfer is especially strange because it occurred during a time of relatively stable climate. No ice sheets were melting during this interval: the last remnants of the great Canadian ice sheet had melted by 7,000 years ago (see Figure 3-6), and the ice sheets on Greenland and Antarctica remained relatively stable during this time. Nor was there any major change in global temperature: regions north of the Arctic

Circle (66.5°N) cooled somewhat during summer because of the decrease in solar radiation (see Chapter 1, Figure 1-10), but temperatures in the middle latitudes and the tropics didn't change much. So why would so much CO_2 have been on the move during a relatively stable interglacial climate?

Comparison to CO_2 Trends in Previous Interglaciations

For another perspective on this problem, we can again focus on the early parts of previous interglaciations to see whether their CO_2 trends were similar to the current one (Figure 3-7). The early parts of the records generally look similar. The CO_2 trends all begin with low values late in the prior glaciations, and they all show CO_2 increases

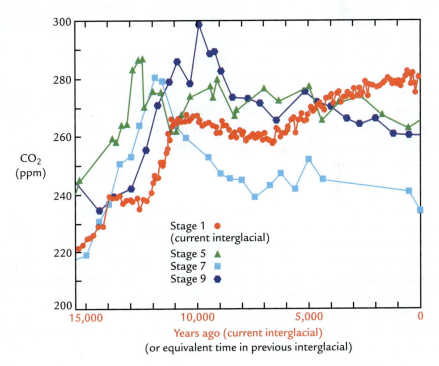

FIGURE 3-7 Trends in atmospheric CO_2 concentrations during the current interglaciation (Stage 1) and the three previous interglaciations. *[Adapted from EPICA Community Members, "Eight Glacial Cycles from an Antarctic Ice Core," Nature 429 (2004): 623-628.]*

similar to the one between 15,000 and 11,000 years ago. Subsequently, CO_2 concentrations reached peaks early in the three previous interglaciations and then began to drop, again similar to the trend in the current interglaciation.

But the CO_2 trends in the later parts of the previous interglaciations are different from the current one. In all three cases, CO_2 concentrations continued to drop slowly throughout the rest of the early interglacial intervals. Not one of these previous interglacial intervals shows a CO_2 increase like the large and steady rise that has occurred during the last 7,000 years. Does this pattern sound familiar? As was the case for methane, the CO_2 trend during the last several thousand years of the current interglaciation has clearly been going the wrong way.

The downward trends that had begun early in the three previous interglaciations are consistent with the long-term link between CO_2 and ice volume (see Figure 3-2). As CO_2 concentrations began to fall, evidence from marine sediments indicates that ice sheets were beginning to grow. These trends match the expected pattern of increasing carbon storage in the deep sea as ice sheets grow.

But that link broke down during the last 7,000 years: CO_2 concentrations rose without any accompanying decrease in global ice volume. Somehow, the CO_2 concentration in the atmosphere increased by almost one quarter of the typical 90-ppm difference between glacial CO_2 minima and interglacial CO_2 maxima without any help from the ice–CO_2 feedback processes that had been so important for the previous several hundred thousand years.

Summary

The main conclusion we can draw from these comparisons is the same one we found for methane: something unexpected, something that had not occurred before, began to happen during the current interglaciation, in this case around 7,000 years ago. CO_2 concentrations trended upward instead of down because some new factor was altering the natural operation of the climate system. Once again, a possible candidate to explain the "wrong-way" CO_2 trend in the last 7,000 years is the emergence of early farming and agricultural practices, which will be explored in the next part of this book.

Additional Resources

COHMAP Project Members. "Climate Changes of the Last 18,000 Years: Observations and Model Simulations." *Science* 241 (1988): 1043–1052.

EPICA Community Members. "Eight Glacial Cycles from an Antarctic Ice Core." *Nature* 429 (2004): 623–628.

Imbrie, J. "A Theoretical Framework for the Pleistocene Ice Ages." *Journal of the Geological Society of London* 142 (1985): 417–432.

Imbrie, J., and K. P. Imbrie. *Ice Ages: Solving the Mystery.* London: Macmillan, 1985.

Kutzbach, J. E., and P. J. Guetter. "The Influence of Changing Boundary Conditions on Climate Simulations for the Past 18,000 Years." *Journal of the Atmospheric Sciences* 43 (1986): 1726–1759.

Petit, J. R., et al. "Climate and Atmospheric History of the Past 420,000 Years from the Vostok Core, Antarctica." *Nature* 399 (1999): 429–436.

Part 1 Summary

We live in a world of interglacial warmth that came into existence nearly 10,000 years ago, near the end of an interval of rapid "deglacial" melting of northern hemisphere ice sheets that began 17,000 years ago. Similar cold-to-warm (glacial-to-interglacial) transitions had happened previously at intervals of about 100,000 years for almost a million years. These natural transitions resulted from changes in Earth's orbit that resulted in greater summer radiation from the Sun at high northern latitudes and melted the snow and ice.

Two important greenhouse gases in the atmosphere—carbon dioxide and methane—also played a role in the ice melting. Although natural variations in the concentrations of these gases result from entirely different processes within Earth's climate system, both the CO_2 and CH_4 levels rose during the most recent deglaciation and reached peak values late in the interval of ice melting.

During three previous warm interglacial intervals, CO_2 and CH_4 concentrations then began to fall even before the last ice remnants had melted, nudging Earth's climate toward the next glacial cycle. During the current interglaciation, these gas concentrations initially followed the same downward trends, but then reversed direction and increased. These upward trends began near 7,000 years ago for CO_2, and near 5,000 years ago for CH_4.

This mismatch between the CO_2 and CH_4 trends in previous interglaciations (Figure 1s-1) and those in the current one poses an intriguing mystery: Why did the gas concentrations reverse direction and rise when they should have fallen? Because the natural concentrations of these two gases are controlled by completely different processes within the climate system, the fact that both follow similarly anomalous upward trends actually doubles the mystery. Why would the factors that control both

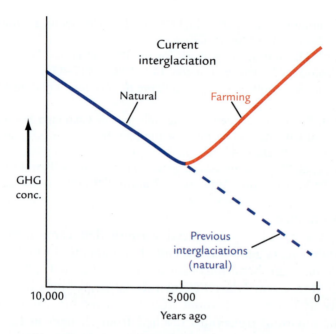

FIGURE 1s-1 The wrong-way greenhouse-gas rise during the present interglaciation (red) differs from the natural downward trends during previous interglaciations (blue), perhaps because of agriculture.

CO_2 and CH_4 behave in such an anomalous way? The next section (Part 2) explores a possible clue: our hunter-gatherer ancestors discovered agriculture nearly 10,000 years ago, and subsequently carried this new way of life across all the arable regions of Earth's continents.

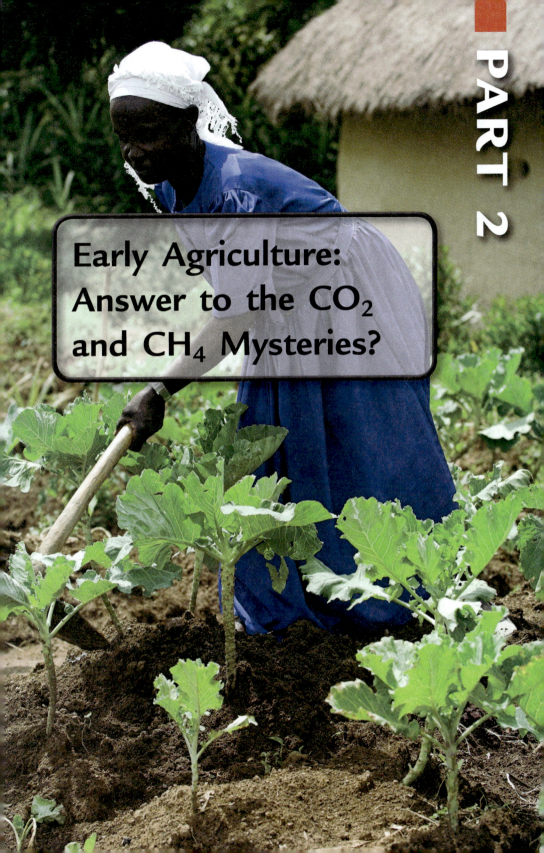

Early Agriculture:
Answer to the CO$_2$
and CH$_4$ Mysteries?

During the last few thousand years when the "wrong-way" increases in atmospheric carbon dioxide and methane concentrations were occurring (see Part 1), a momentous new discovery—agriculture—was spreading across all the continents except for Antarctica and Australia. This new trend is potentially important in explaining the anomalous greenhouse-gas increases because agriculture is a source of both CO_2 (from **deforestation,** including both the cutting and burning of forests) and CH_4 (from rice irrigation, livestock tending, and incomplete burning of grass and scrub brush).

Scientists have long followed the saying that "correlation is not causality." That is, the existence of a correlation is not sufficient on its own to prove a cause-and-effect link. Still, the correlation in time between the wrong-way gas trends and the spread of agricultural practices that emitted those gases is an intriguing observation that deserves to be explored.

Such an exploration leads to questions that span a vital part of human history and prehistory. Where and how did agriculture originate on the various continents? Once begun, how and why did it spread? More specifically, when did people in different regions first begin to clear forests, tend livestock, construct irrigated rice paddies, and otherwise transform landscapes? As these transformations occurred, how did the methods of farming change?

Even the oldest written summaries of human activities only take us back a little over 3,000 years, although older records exist from Sumeria in modern-day Iraq. These records, initially gathered mostly for purposes of food distribution and tax collection, suggest that by 2,500 to 2,000 years ago Earth's human population had grown to an estimated 250–300 million people, a small fraction of the 7 billion people alive today, yet hardly a trivial number. Unfortunately, we have very few reliable national-level surveys from prior to fifty years ago that explore how much forest had been cleared and how much land had come under cultivation or been put to use as pasture for livestock.

Within these constraints, it might seem that most of the 10,000-year or longer history of agriculture would be beyond our understanding, but this is not the case. Scientists working in an array of disciplines have filled in many of the broad outlines of this world-changing story. Much of this global-scale panorama has been gradually assembled over decades by the slow, hard, "dirty boots" work of thousands of researchers who went out into the field and carried out studies in archeology, ecology, anthropology, geology, botany, and related areas.

Because these efforts span such a wide range of disciplines, no one term can possibly encompass them all. The one I prefer is *land-use archeology*. Decades ago, most archeologists concentrated on the buildings, monuments, and graves of wealthy ruling classes in advanced civilizations, that tiny fraction of ancient people who led privileged lives completely unlike most of humanity. In contrast, the evolving field of land-use archeology concentrates on basic day-to-day activities such as clearing land, building homesteads, planting crops, tending livestock, and creating and using tools. In this way, land-use archeology assembles an overview of the cumulative environmental effects of the many millions of "common people" who were the ones that slowly transformed the surface of much of this planet.

For example, analyses of pollen grains deposited in lakes and ponds provide insights into the gradual replacement of natural vegetation (mostly forest) by crops and weeds associated with clearance and farming. Analyses of plant and crop remains by archeobotanists show how domesticated plants evolved from natural ones, with many recent insights gained from genetic information. Analyses of silts and clays deposited in lakes, riverbanks, and deltas tell us how sediments eroded from the cleared land and were carried away by water and wind.

Among eleven centers of plant and livestock domestication (Figure 2i-1), three were particularly important: the **Fertile Crescent** region

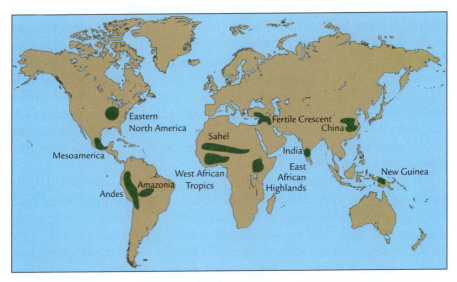

FIGURE 2i-1 Centers where agriculture originated. *[Adapted from P. Bellwood,* First Farmers: The Origins of Agriculture *(Oxford: Blackwell, 2004), and M. D. Purugganan and D. F. Fuller, "The Nature of Selection During Plant Domestication,"* Nature *457 (2009): 843–848, doi:10.1038/nature07895.]*

of southwest Asia (modern-day Turkey, Syria, Iraq, Jordan, and Israel), China, and Mesoamerica (modern-day Mexico). Livestock domesticated in these areas thousands of years ago provide beef, lamb, pork, milk, cheese, and yogurt today. Among the dozens of wild plants converted to domesticated crops in these regions were three especially important grains that have nourished humans ever since: wheat in the Fertile Crescent, rice in China, and corn (*maize*) in Mexico. These grains give us the cereals we eat today for breakfast, bread for lunchtime sandwiches, and corn on the cob for evening meals. Other early-domesticated grain crops include the barley that we use in our soups, and the rye and hops that we distill and ferment to make our evening drinks.

All in all, fifteen crops first domesticated many thousands of years ago in these various ancient agricultural centers account even today for most food consumption on this planet: the grains—wheat, corn, rice, barley, and sorghum; the roots—potato, manioc, and sweet potato; the green vegetables—soybeans, squash, beans, and lentils; the sugar sources—cane and beets; and bananas (a tree fruit). Many of the earliest forms of these crops were smaller than modern varieties, but they are still easily recognizable as the same plant.

The interval 12,000 to 10,000 years ago when agriculture first began to appear is called the **Neolithic Agricultural Revolution** (often shortened to Neolithic Revolution). *Neolithic* indicates that humans were still using tools made of stone (*lithic*). *Neo* (new) suggests that the tools were more delicately shaped and polished than the crudely chipped tools of earlier times. The first bronze tools do not appear until the Bronze Age, 5,000 years ago, and the first iron tools appear in the Iron Age, 3,000 to 2,500 years ago (Figure 2i-2). For at least 100,000 years before the Neolithic Revolution, Stone-Age people, mostly members of our own species, had been living off what Nature provided, by hunting wild animals and birds; gathering wild cereals, seeds, nuts, berries, and roots; and catching fish.

It is estimated that a few million people lived this way until 10,000 years ago. One way demographers arrive at this estimate is to start with the reasonable value of roughly 250–300 million people at the start of the historical era 2,000 years ago and project back in time using best-guess assumptions about how fast populations grew. One such estimate is shown in Figure 2i-2. Other estimates rely on transforming anthropological studies of the density of peoples who still live as hunter-gatherers today into estimates of past populations.

The Neolithic Revolution set off an explosive increase in population that started humanity on its path toward more than 7 billion people

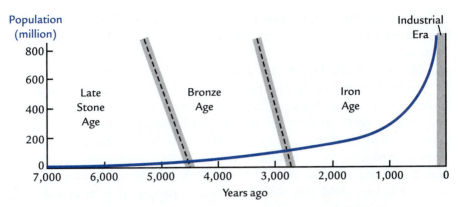

FIGURE 2i-2 Rising global population trend during the late Stone Age, Bronze Age, and Iron Age. *[Adapted from C. McEvedy and R. Jones,* Atlas of World Population History *(New York: Penguin, 1978).]*

today. During the last 10,000 years, farmers have dispersed newly domesticated crops and livestock across the low- and mid-latitudes of the continents, in some cases also introducing new cultures and languages. Over time, the greater reliability of food from farming and the resulting growth of large populations led to urban civilizations and writing (in part to keep track of food storage and distribution in increasingly large and complex societies). In a real sense, this ancient revolution was the key point of departure that created the human world we know today.

Yet the reason this agricultural revolution began is not so obvious as it might seem. On the positive side, agriculture made it possible for people to live a settled life by providing abundant food during most warm seasons and crop surpluses during favorable years to carry forward into the unproductive winters and drought years that occasionally followed. The availability of food during hard times probably helped to reduce natural mortality levels enough so that populations could begin to grow.

But farming also required hard work. In many regions, clearing forests combined with planting, weeding, and harvesting crops took more time than simply going out and foraging. Also, studies of human bones from this time suggest that in some areas the nutritional value of relatively starchy cereal crops may have been inferior to the nutritious mix of foods consumed by those living as hunter-gatherers. In addition, the constant bending and stooping required for working in the fields and grinding cereal grains produced many deformities in the bones of early farmers. If farming was harder work, and if some crops were less nutritious than mixtures of wild foods, why did people make the momentous transition to agriculture, and why did populations increase?

One important factor may have been the effect on female fertility of being able to live in one place during the brief child-bearing and child-rearing years. Because hunter-gatherer people moved constantly, they had to carry all their possessions with them—tools, food supplies, parts of their dwellings, and their infant children. These multiple burdens limited how many infants could be carried during each move, and this constraint favored spacing births widely to avoid having to carry more than one child at a time. In contrast, people who farmed did not have this burden because they remained in one place; they could have children more frequently. Also, the availability of cereal grains and milk from livestock may have favored the survival of infants who could be weaned early on soft foods (cereal mush or porridge). Whatever the mix of reasons, human populations that had long been held in check began to increase explosively as agriculture spread.

Another mystery is why agriculture first appeared in several places at approximately the same time: nearly 11,000 years ago in the Fertile Crescent, and soon after 10,000 years ago in northern China and parts of the Americas. Some scientists have suggested that these near-simultaneous transformations were linked to climate. One theory is that the shift from cold, dry, constantly oscillating late-glacial climates to warmer, wetter, and more stable interglacial conditions made those regions more favorable for agriculture. Another (nearly opposite) theory is that an abrupt reversal to cooler and drier climates between 13,000 and 11,500 years ago (the brief Younger Dryas event noted in Chapter 2) forced hunter-gatherer people to become more innovative about finding alternative food sources. Neither of these ideas explains why full-fledged agriculture didn't appear for another 5,000 years in other regions.

Well before the last 10,000 years, our species had already developed a number of intellectual and technological skills that enabled them to better adapt to the varying environments in which they lived. One of those skills was a deep first-hand familiarity with the nearby plant and animal life, both of which were day-to-day sources of food. Perhaps the simplest explanation for the (several) independent origins of agriculture is that this growing knowledge, along with a growing openness to innovation, predisposed these widely separated groups to take similar steps into a new way of life in regions where conditions were favorable.

The four chapters in Part 2 focus on the large-scale changes in landscapes that occurred as agriculture spread across the major continents.

Each chapter will show that the greatest spread of agriculture occurred during the interval of "wrong-way" trends in the two greenhouse gases summarized in Part 1: the CO_2 reversal and rise that began 7,000 years ago, and the CH_4 reversal and rise that began 5,000 years ago.

Additional Resources

Adams, R. M. *The Evolution of Urban Society: Early Mesopotamia and Prehispanic Mexico.* London: Aldine, 1966.

Bellwood, P. *First Farmers: The Origins of Agricultural Societies.* Oxford: Blackwell, 2004.

Diamond, J. *Guns, Germs, and Steel: The Fates of Human Societies.* New York: W. W. Norton, 1999.

Fagan, B. *The Long Summer: How Climate Changed Civilization.* New York: Basic Books, 2003.

Fuller, D. Q., R. G. Allaby, and C. Stevens. "Domestication as Innovation: The Entanglement of Techniques, Technology, and Chance in the Domestication of Cereal Crops." *World Archeology* 42 (2010): 13–28. doi: 10/1080/00438240903429680.

Kirch, P. V. "Archaeology and Global Change: The Holocene Record." *Annual Review of Environment and Resources* 30 (2005): 409–40. doi.10.1146/annurev.energy.29.102403.140700.

Martin, P. S. "Prehistoric Overkill: The Global Model," in *Quaternary Extinctions: A Prehistoric Revolution,* edited by P. S. Martin and R. G. Klein. Tucson: University of Arizona Press, 1984.

McEvedy, C., and R. Jones. *Atlas of World Population History.* New York: Penguin Books, 1978.

Roberts, N. *The Holocene: An Environmental History.* Oxford: Wiley-Blackwell, 1998.

Simmons, I. G. *Changing the Face of the Earth: Culture, Environment, History.* Oxford: Wiley-Blackwell, 1996.

Thomas, W. L. Jr. *Man's Role in Changing the Face of the Earth.* 2 vols. Chicago: University of Chicago Press, 1956.

Tudge, C. *The Time Before History: 5 Million Years of Human Impact.* New York: Touchstone, 1997.

Wells, S. *The Journey of Man: A Genetic Odyssey.* New York: Random House. 2003.

Williams, M. A. *Deforesting the Earth: From Prehistory to Global Crisis.* Chicago: University of Chicago Press, 2003.

CHAPTER 4

THE FERTILE CRESCENT AND EUROPE

European agriculture resulted from innovations that began more than 10,000 years ago, and not in Europe. Farming originated east of the Mediterranean Sea in the Fertile Crescent region of southwest Asia. There, ingenious groups of people domesticated many of the crops humanity relies on today for food (wheat, barley, peas, chickpeas, and lentils), as well as the major livestock we raise today (cattle, sheep, goats, and pigs) (Table 4-1). The beginning of the spread of agriculture from this place of origin can be traced from field studies in archeology and related disciplines, and an increasingly detailed history of agriculture since 2,000 years ago can be reconstructed from written documents.

Europe is a small, originally forested continent separated from the much larger mass of Asia by the Ural Mountains to the east (see Figure 4-1). South from a thin strip of Arctic **tundra** (grasses and low-lying shrubs), the forests of Europe range from high-latitude **conifer forests** (such as spruce and pine) in the Scandinavian north to broad-leafed, deciduous **hardwood forests** (such as oak, ash, elm, maple, and beech) over most of central Europe, and then to a band of distinctive Mediterranean vegetation (including cedars, pines, and evergreen oaks) in the south. The semi-arid **steppes** in southeast Europe are grassland regions that lack forests, except along river valleys.

TABLE 4-1 Crop and Livestock Domestication in the Fertile Crescent

Crops	Time Domesticated
Rye	13,000+/– years ago
Wheat (emmer)	11,000–6,000 years ago
Wheat (einkorn)	11,000–6,000 years ago
Barley	11,000–6,000 years ago
Peas	11,000–6,000 years ago
Lentils	11,000–6,000 years ago
Chickpeas	11,000–6,000 years ago
Figs	By 6,500 years ago
Dates	By 6,500 years ago
Olives	By 6,000 years ago
Oats (Europe)	By 5,500 years ago

Livestock	Time Domesticated
Goats	10,500–10,000 years ago
Sheep	10,500–10,000 years ago
Pigs	10,500–8,000 years ago
Cattle	By 8,500 years ago

The Early Prehistorical Record: Hunting and Gathering

DNA studies have traced the geographic movements of *Homo sapiens* through time. The Y chromosome (in males) slowly undergoes subtle genetic mutations that mark the divergence of different groups from their common ancestors and from each other. Because the rate of genetic mutation is constant, these DNA changes provide a "clock" that can be used to time these divergences and migrations. Previous studies estimated migration patterns of early humans using similarities and differences between languages spoken by peoples in different parts of Europe. DNA studies have both confirmed and extended the results of those language studies.

DNA evidence indicates that Neanderthals had been present in Europe for tens of thousands of years prior to the arrival of Cro-Magnon people (anatomically modern humans) some 45,000 to 35,000 years ago. Cro-Magnon people appear to have come from southwest Asia, probably

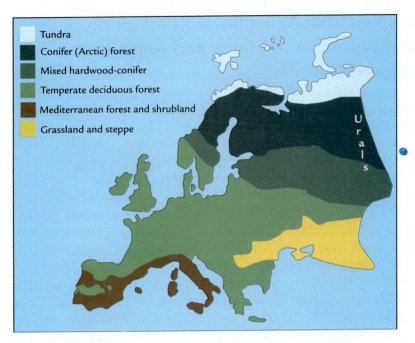

FIGURE 4-1 Major vegetation covering Europe to the Ural Mountains. *[Adapted from R. L. Smith and T. M. Smith,* Elements of Ecology *(Menlo Park: Benjamin Cummings, 1998).]*

following herds of wild game on grassy steppes (grasslands) that in glacial times extended farther west than today (Figure 4-1).

How the Cro-Magnons displaced the Neanderthals is not clear. If it happened through conflict, the Cro-Magnon people must have had superior weapons and fighting strategies to defeat the much stockier and more muscular Neanderthals. A more likely explanation is that Cro-Magnons had superior communication skills that gave them tactical advantages, not just in conflicts but also in other activities, such as pursuing and killing wild game. With these skills, they could have outcompeted the Neanderthals by killing more of the game both groups relied on to supplement gathered food. Other social and technological skills may also have increased the survival rates of Cro-Magnons. Whatever the reason, the Neanderthals were extinct by 25,000 years ago.

The hunter-gatherer Cro-Magnon people lived by hunting wild game (deer, elk, wild ox, gazelles, and smaller animals), by gathering berries, seeds, nuts, and roots (tubers), and by fishing in rivers, lakes, and along coastlines. These early hunting activities had no drastic effect on most wild game species in Europe, although a few became extinct.

The Neolithic Agricultural Revolution: Domestication of Plants and Animals

The Neolithic Revolution started in the Fertile Crescent around 12,000 years ago (Figure 4-2). Prior to that time, all people on Earth lived by hunting, gathering, and fishing, but Fertile Crescent peoples began to domesticate crops and livestock, and by 8,000 years ago, they had become dependent on agriculture for most of their food.

Plant domestication arose because hunter-gatherer people had long picked and eaten seeds from natural grasses (grains) as an important part of their diet. The Fertile Crescent was particularly favorable to this kind of foraging, because it had fertile soils and reasonably reliable winter rainfall, and because it was richer in wild annual plants that produced larger seeds than anywhere else in the world. Eventually, when people in the eastern Mediterranean noticed that plants sprouted from uneaten or undigested seeds in nutrient-rich fire ash and in piles of human waste, they began scattering the seeds on the ground and then, later, planting them in small plots. By doing so, they became farmers. At first, no special technology was required; they used wooden digging sticks (called **dibbles**). To save time and energy, they chose the largest of the available seeds, thus starting a process of artificial selection (**domestication**).

In the wild, many seeds aid their own dispersal by **shattering**—bursting from the parent plant when lightly brushed against or when blown by winds. When people gathered seeds from the wild or from their own small garden plots, they naturally chose the ones that were still on the plant,

FIGURE 4-2 The Fertile Crescent region in southwest Asia. *[Adapted from J. Diamond, Guns, Germs, and Steel: The Fates of Human Societies (New York: W. W. Norton, 1997).]*

FIGURE 4-3 Wheat domesticated in the
Fertile Crescent. [Robert L. Smith.]

rather than those that had already shattered
and dispersed. These ongoing choices on
the part of early farmers resulted in
cultivation of varieties of plants that kept
more seeds than they dispersed.

As the selected plants gradually became more resistant to shattering, early
farmers harvested them with **sickles.**
Then they had to thresh and winnow the
grain in order to separate the wheat from
the chaff, adding considerably to the time
and labor required to bring in crops. As this process unfolded, the
grains became increasingly unable to self-sow in the wild and were
completely dependent on propagation by humans. Scientists once
thought that the domestication process for each type of grain occurred
quickly (within a century), but recent archeobotanical research suggests
that it took several thousand years for farmers to eliminate the shattering behavior completely. Domestication of wheat and barley started
around 11,000 to 9,000 years ago but wasn't fully complete until 7,000
to 6,000 years ago.

The most active center of plant domestication was a broad arc in
the Fertile Crescent that spanned northern Syria and southeastern Turkey,
as well as parts of Iraq, Jordan, and Israel (see Figure 4-2). Archeobotanists
have traced discrete areas where the various Fertile Crescent crops were
domesticated. Wheat and
barley, two of the major food
sources in the world, were
among the first to be domesticated; rye and millet, along
with peas, chickpeas, and
lentils came not long after
(Figures 4-3 and 4-4). Flax
was also domesticated as a
source of linseed oil and of

FIGURE 4-4 Chickpeas
domesticated in the Fertile
Crescent. [Robert L. Smith.]

FIGURE 4-5 Reconstruction of an early agricultural sickle. *[Erich Lessing/Art Resource, NY.]*

fiber for linen. The crop food sources were supplemented with acorns, blackberries, crab apples, and nuts gathered in the wild. The demands of farming also led to the use of tools that had been developed earlier such as sickles for cutting wild grain (Figure 4-5) and mortars and pestles for grinding it.

Wild animals were domesticated during the same interval as early crops were. Gazelles had long been a major food source, but their numbers had been greatly reduced by hunting, as had those of larger game such as rhinoceros, red deer, horses, and aurochs (large predecessors of domesticated cattle, now extinct). The average size of the remaining gazelles became smaller, indicating that people were gradually being forced to hunt juvenile populations. In addition, remains of smaller creatures like hares, fish, birds, and even land snails became more common in archeological deposits.

As with wild plants, the Fertile Crescent was fortunate to be home to several kinds of wild animals with the size and temperament needed for domestication. After many millennia of hunting wild animals or driving them into traps for killing, people began to use food to attract

FIGURE 4-6 Livestock domesticated in the Fertile Crescent. *[Stefan Boness/Panos Pictures.]*

animals to areas near their settlements, and by degrees began to constrain their movement and then tend them in herds. Sheep, goats, and pigs were domesticated by 10,500 to 9,000 years ago, and cattle by 8,000 years ago (Figure 4-6). Cattle were domesticated from aurochs, and pigs from wild boar. Domestication of livestock gradually produced animals that were smaller (and more docile) than their wild predecessors.

Proof of domestication comes from archeological sites that contain bones only from male and older female animals, an indication that younger females were being spared for breeding purposes. Genetic data show many separate sites of domestication of different strains of livestock and subsequent dispersal to other regions. In time, the meat, butter, cheese, milk, and yogurt from domesticated livestock replaced wild game as the major source of protein.

The fact that these two momentous processes—the domestication of crops and livestock—happened in this one relatively small area is obviously linked to the fortuitous presence of wild plants and animals of the right size and temperament for domestication. But why did so much domestication occur in such a (relatively) short period of time?

In the case of livestock, the gradual over hunting of wild animals could be a common region-wide explanation. Also, it seems likely that after one or two groups of people initially came up with these new ideas, the concepts then spread to neighboring groups, who adapted them to the wild plants and animals available in their particular areas.

The wide variety of domesticated crops and livestock provided surplus food for years when harvests were poor, but also constrained people to one location year-round in order to plant and tend their crops, and guard their food stores and livestock. As a result, small settlements, villages, had developed by 10,500 to 9,000 years ago, well before crops were fully domesticated. Most regions in the Fertile Crescent had become fully agricultural economies by this time, even though the Bronze Age and Iron Age lay thousands of years in the future, and the tools they provided—metal plows and hoes—were not yet available.

The Spread of Fertile Crescent Agriculture into Europe

The extent of agriculture through time can be reconstructed through **radiocarbon dating** of the remains of domesticated crops and livestock in archeological deposits. Prior to 10,000 years ago, the Fertile Crescent "package" of crops and livestock was confined to just this one relatively limited part of southwestern Asia (Figure 4-7). By 9,000 years ago, evidence of the Fertile Crescent crops can be found in the Balkan countries (modern-day Greece and Bulgaria), and signs of forest clearance and environmental degradation begin to appear in the archeological record. It seems unlikely, however, that the deforestation of this semi-arid region could have emitted much CO_2 to the atmosphere.

Then, near 7,500 years ago, remains of Fertile Crescent crops suddenly appear in archeological deposits across much of Europe. Because most of Europe was heavily forested, growing these crops generally required cutting down trees to let sunlight in. The earliest European settlements are found along river valleys where floods deposited nutrient-rich, water-retentive soils, and in nearby areas where strong glacial-age winds had left silt-sized deposits called **loess**. The nutrient-rich and easily tilled loess soils were intensively farmed year after year. At about this time, oats were domesticated somewhere in Europe by people growing Fertile Crescent crops. Cattle and pigs became the most common kind of livestock in northern Europe, while sheep and goats were favored in the south.

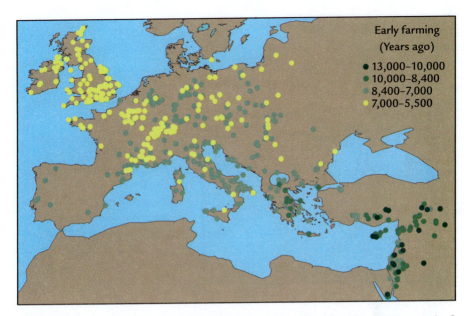

FIGURE 4-7 Radiocarbon-dated archeological sites showing the early spread of Fertile Crescent crops across Europe. *[Adapted from D. Zohary and M. Hopf, Domestication of Plants in the Old World: The Origin and Spread of Cultivated Plants in West Asia, Europe, and the Nile Valley, 3rd ed. (Oxford: Oxford University Press, 2000); updated by C. S. M. Turney and H. Brown, "Catastrophic Early Holocene Sea Level Rise, Human Migration, and the Neolithic Transition in Europe,"* Quaternary Science Reviews 27 (2007): 2036–2041.]

Although this was still the Neolithic period, trees were felled using sophisticated, polished flint axes by 9,000 years ago (Figure 4-8). Earlier axes had been used for woodworking, but these new axes were sturdy enough to cut notches around the circumference of trees to stop the flow of sap in the outer layers. After this **girdling**, trees dropped dead branches and limbs to the ground, and farmers set fire to the debris during dry spells. Based on modern-day tests of these efficient flint axes, remarkably little time would have been needed for one farmer to girdle acres of trees.

FIGURE 4-8 Polished flint axes used to girdle trees. *[Ran Barkai, Tel-Aviv University, Israel.]*

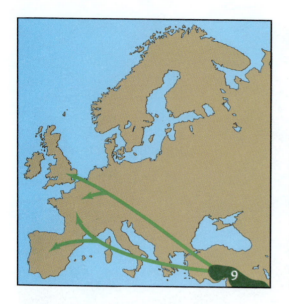

FIGURE 4-9 Two major pathways of Fertile Crescent agriculture penetration into Europe. The number *9* refers to the time (9,000 years ago) that crops began to spread beyond the Fertile Crescent region. [*Adapted from P. Bellwood, First Farmers: The Origins of Agricultural Societies (Oxford: Blackwell, 2004).*]

This new way of life was spread across Europe both by people moving west across the landscape and by existing hunter-gatherer people adopting the new agricultural concepts. Studies of similarities in language and in genetic material support the archeological evidence by showing two main pathways for the westward spread of agriculture (Figure 4-9). Agriculture propagated slightly earlier along a southern route following the northern coastlines of the Mediterranean Sea, reaching Cyprus and the Greek islands and then moving westward into Italy and Spain. A slightly later northern route followed the Danube River valley northwest into Germany, and then spread west into the Paris Basin of France and into Britain and Ireland.

Scientists have long debated whether the migration of people or the spread of new ideas was more important to the spread of agriculture. Current evidence suggests that both processes were involved, but the movement of farming people was probably more important. In any case, adoption of the Fertile Crescent agricultural package occurred quickly in each region of Europe. Radiocarbon-dated lake sediments show the abrupt appearance of European weeds typical of forest clearance: grasses, plantain, nettles, and dock.

As noted earlier, many of the earliest European farmers settled in areas of rich loess soils that remained fertile for many years, but later arrivals farmed areas where soil fertility was depleted more quickly. Initially, scientists thought that these early farmers moved around often, farming one area for a few years until the soil fertility ran low, and then moving their dwellings to another area, where new plots had to be cleared. Now, it appears more likely that many farming families stayed in the same homes for long intervals of time, but repeatedly shifted their cultivation activities among different plots, letting some land lie abandoned. After two-to-three decades, shrubs and immature trees had grown on the abandoned plots, and the soil had regained much of its original fertility. At that point, the vegetation on these plots was set afire, and the soils were farmed for a few years until their fertility again ran low.

This kind of shifting cultivation is called **forest fallow** or "long fallow" farming, terms that refer to the long intervals during which abandoned plots are left unused ("lying **fallow**"). Also called "slash and burn" agriculture, this method used a large amount of land per family because of the constant shifting from plot to plot. Because the population density in Europe at this time was still low, many farmers had few or no neighbors (except within their immediate family group, or clan) and thus probably had little sense of needing to set limits on their use of the land.

Livestock also played a role in forest clearance. Some cleared land was cultivated for a while and then converted to pasture, where browsing animals ate not only the grass but also any emerging shoots of trees and shrubs. Because of this browsing, forests could not regenerate in these areas. Livestock were pastured in lower terrain during the winter season and taken to higher elevations during summer.

Before long, substantial clearance had begun even in relatively remote areas, such as villages high in Alpine valleys. Evidence for the extent of Alpine clearance is found in radiocarbon-dated wood preserved in the ruins of former dwellings. The wood in the earliest village structures comes from trees that grew in valleys nearby, but wood used in later buildings comes from trees typical of more remote, high-elevation areas. This evidence suggests that the nearby forests in the lower reaches of the valleys had run out of desirable trees.

Anthropologist Susan Gregg attempted to estimate how much land would have been used by a typical small village about 6,000 years ago in central Europe (modern-day Germany). She assumed that the six families (thirty people) living in the village derived their food from small plots of cultivated crops, from livestock fed in large pastures and winter fodder cut from hayfields, and from pigs allowed to roam freely in the nearby woods and eat acorns and other food. She calculated the wood requirements for fuel and buildings by assuming that wood lots were cut on a 30-year rotation. And she estimated the small area that would have been taken up by roads and dwellings. By her estimates, a village would have needed a total of just under 120 hectares (about 265 acres), or a little less than 4 hectares (8–9 acres) per person. Early farming required a surprisingly large amount of land.

By 5,500 years ago, archeological records document many additional sites with Fertile Crescent crops that reveal a still greater spread of agriculture, particularly in northwestern and northern parts of Europe (see Figure 4-7). By this time, agriculture was underway in every part of Europe where it is practiced today. Although most of these sites were probably still small clearings in the woods, the evidence suggests that the cumulative amount of early deforestation in Europe could have played a role in the wrong-way CO_2 trend. The interval of rapid agricultural expansion between 7,500 and 5,500 years ago brackets the time during which the CO_2 trend reversed direction from a downward to an upward trajectory (see Chapter 3, Figure 3-7).

By 5,000 years ago, just prior to the start of the European Bronze Age, both southwest Asia and Mediterranean Europe had been substantially transformed by humans. In the Fertile Crescent, agricultural villages were turning into towns, and the world's first urban area—Sumer—appeared along the Tigris and Euphrates rivers. Urbanization was accompanied by the world's first political state, the earliest form of writing, and the invention of the wheel. In the Fertile Crescent, farmers began planting orchards of olives, figs, dates, grapes, and pomegranates, which were later adopted in Mediterranean Europe (see Table 4-1). Horses that had been domesticated around 6,000 years ago in the steppes of central Asia north of the Black Sea appeared in Europe. Ox-drawn wooden plows came into use.

Britain: Case Example of Prehistorical Agriculture

Between 5,500 years ago and the start of the early historical era 2,000 years ago, farming populations and their effects on the landscape of Europe grew. Britain serves as a useful example of the dramatic changes in the European landscape during this interval, because its land-use patterns have been intensively studied in recent decades. The timing of land-use changes in Britain seems likely to be intermediate between the earlier transformation of the classical Mediterranean civilizations in Greece and Rome, and the later changes in more remote regions like western Russia, and thus more representative of Europe at large.

Archeological evidence places the first substantial opening of Britain's forests between 6,500 and 5,000 years ago, with plowing of lowlands and

FIGURE 4-10 Early wooden ard plow from Europe. Draft animals pulled plows with the handles at the left; the sharp stick at the right cut narrow lines in grassy sod. [Robert L. Smith.]

some farming even on less fertile hillsides. The evidence for increased land use includes radiocarbon-dated crop and livestock remains found in lake sediments (see Figure 4-7), large increases in the number of radiocarbon-dated archeological sites, changes from predominantly forest pollen to pollen from weeds and open land, and the remains of wooden hoes and flint axes. For some time previously, farmers had been using primitive **ard plows** (also called "scratch plows") cut from the wood of strong trees like oaks (Figure 4-10). These plows, simple wooden sticks dragged through relatively light (loose) soils by a single draft animal, incised criss-cross "X" patterns in layers of sod so that they could be easily turned with wooden shovels. Remains of these X patterns are still preserved in some areas under mounds of soil heaped around ancient burial sites.

By 5,000 years ago, deforestation had begun to destabilize and erode hillsides. Small increases in the accumulation of mud in lakes and rivers indicate that disturbed soils were being moved around by water and wind. Indications of such soil disturbance are found in Britain and elsewhere in Europe, including the Rhine Valley of Germany (Figure 4-11). These influxes of mud continued and intensified during subsequent millennia as clearance continued.

Between 4,000 and 3,000 years ago, families began to build low stone walls called **reaves** to mark the boundaries of their fields. The need for walls suggests that land had become less easily available and more valuable, and the walls themselves also provide some idea of the growing extent of land use. Use of the ard plow increased during this interval, and soils in some areas begin to show the first evidence of nutrient depletion and build-up of acidic soils.

Early in the Iron Age, after 3,000 years ago, numerous **hillforts**, large walled enclosures on hilltops, also appeared, seemingly as forerunners of later towns (Figure 4-12). In addition to defensive uses, these structures served as places where farming communities could thresh, sieve, and store grain. People also kept and exchanged livestock at these

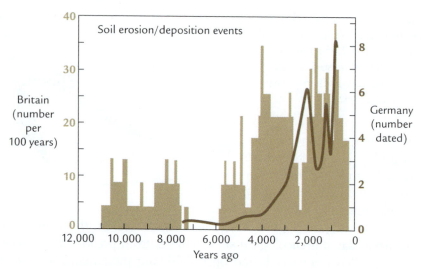

FIGURE 4-11 Increasing sediment influx into European lakes after 5,000 years ago. *[Adapted from M. G. Macklin, "Holocene River Environments in Prehistoric Britain: Human Interaction and Impact,"* Journal of Quaternary Sciences *14 (1999), 521–530; and from A. Lang, "Phases of Soil Erosion-derived Colluviation in the Loess Hills of Southern Germany,"* Catena *51 (2003), 209–223.]*

FIGURE 4-12 Remains of an early hillfort in England. *[Robert Harding Picture Library Ltd./Alamy.]*

sites, some of which were permanently occupied to defend common food sources. Recent tests show that fires set at these hilltop sites are easily visible from nearby hillforts. Perhaps they served as nodes of communication among the early populations.

Also during this period, "round houses" appeared—structures with round supporting walls made of wood or of stone patched with mud and straw ("wattle and daub"), and topped by cone-shaped thatched roofs. These dwellings no longer sat in remote clearings surrounded by forests, but instead were located in open landscapes that had been mostly cleared for agriculture, with only patches of woodlands left. The number of livestock also increased during this interval, with grazing shifting back and forth, from lower pastures that stayed mild during cold winters to higher mountain pastures that were cool during the summer heat.

The Historical Record

Written evidence in Europe begins during classical Greece, with several observers (including Herodotus and Aristotle) commenting on the pervasive hillside deforestation in nearby areas. Lucretius, a poet and philosopher, noted similarly widespread clearance in Rome, consistent with the fact that lumber to build Roman ships had to be brought in from distant sources—the Alps, North Africa, and Spain.

From a population that was probably well under a million people 7,500 years ago, Europe had grown to 30 million by the dawn of the historical era 2,000 years ago. During this same interval, the concentration of CO_2 in the atmosphere had risen (see Chapter 3, Figure 3-7). The release of CO_2 from clearing and burning European forests must have played some role in the rising CO_2 trend. Similarly, the proliferation of livestock as the European population grew would have contributed to the CH_4 increase after 5,000 years ago (see Chapter 2, Figure 2-8).

The interval that began 3,000 years ago and extended through the Roman classical age is regarded by many scientists as the most active phase of forest clearance in Britain's entire history (Figure 4-13). Clearance was most intense at first in lowland Britain toward the south and southeast, and then increased later in higher, less arable areas in the north and northwest. Lake sediments dating to this time period show sharp increases of pollen from weeds, indicating forest clearing. Lake and floodplain sediments in Britain and other parts of western and central Europe record even greater influxes of silts and muds at this time (see Figure 4-11). By this period, many of the remaining woodlands were managed

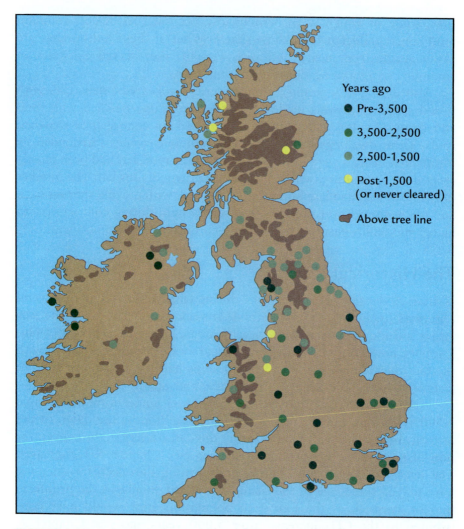

FIGURE 4-13 Archeological sites showing the timing of prehistorical and early historical clearance of forests in Britain and Ireland. *[Adapted from N. Roberts,* The Holocene: An Environmental History *(Oxford: Wiley-Blackwell, 1998).]*

as a supply of fuel wood, fence posts, and rails. Trees were modified through **coppicing**—pruning above ground level to encourage subsequent growth of multiple branches of useful sizes.

Early in the Roman era, Julius Caesar in his *Commentaries on the Gallic War* described a trip to Britain: ". . . the population of [southern Britain] is exceedingly large, the ground thickly studded with homesteads." Much of the modern pattern of rural towns and roads in the

British countryside was established during the Roman era (the years 43–420). The botanist and landscape historian Oliver Rackham estimated a total population in Britain of at least 4 million people by late Roman times, with most of the arable land under cultivation. Early Iron Age plows cut furrows in the soil, aiding farmers by keeping weeds and shrubs from gaining foothold in fallow fields.

Clearance was also pervasive at this time across much of southern, western, and central Europe eastward to at least central Germany. Julius Caesar noted substantial clearance even across the northernmost German fringes of the Roman Empire, and rates of erosion and sedimentation accelerated in the Rhine Valley during this time (see Figure 4-11). After land was cleared, browsing goats and sheep kept the pastures clear, especially in Mediterranean areas. Goats and sheep are particularly hard on the land, eating every bit of vegetation available. To combat the resulting erosion, farmers terraced steeper hillsides and planted orchards and vineyards, but the problem grew worse after the year 400 when the central organizing authority of the Roman Empire collapsed. Coastal towns and cities had to be repeatedly relocated seaward as their ports filled up with silty mud caused by erosion. The original harbor at Troy (in modern-day Turkey) eventually had to be moved almost 4 kilometers seaward from its original location (Figure 4-14).

Between the years 200 and 600, the population of Europe fell by some 10 million people (about a 40% drop). A major factor in this decline was the increased incidence of disease, including a massive outbreak of bubonic plague in the year 540. The worst population losses were in the Mediterranean south, with Germanic populations to the north less affected. As a result, the center of population in Europe shifted toward the north,

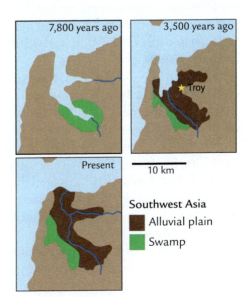

FIGURE 4-14 Seaward relocation of the city of Troy because of the accumulation of silty mud. [*Adapted from N. Roberts,* The Holocene: An Environmental History *(Oxford: Wiley-Blackwell, 1998), based on J. C. Kraft, I. Kayan, and O. Erol, "Geomorphic Reconstructions in the Environs of Ancient Troy,"* Science *209 (1980) 776–782.]*

with disastrous effects on the remains of the Roman Empire. Germanic people first raided southward into Rome more than 2,000 years ago, then again in the year 59, and many other times between 200 and 500 as Mediterranean populations fell. Not all of these incursions were tied to warfare; at times, families of farming people simply walked south and settled on abandoned land. Central Asian peoples also raided the Roman Empire.

In Britain, during and after the 540 outbreak of bubonic plague, some land reverted to forest (**reforestation**). During the "dark age" that followed (from 600–1000), little additional deforestation occurred in Britain (or much of the rest of Europe) as population numbers remained depressed for centuries.

In 1086, William the Conqueror ordered a survey of his newly conquered domain (England). The 1089 **Domesday Book** recorded a population of 1.5 million people living in a land where 85% of the forests had been cleared, including more than 95% of the fertile low-lands in the southeast. No tracts of forest more than 6–8 miles across remained, consistent with the general consensus that clearance in Roman times had been nearly complete.

Between 1000 and the middle 1300s, population estimates based on widely available parish registers and household counts (**hearth counts**) surged everywhere in Europe, and England, France, and Spain became nation-states with substantial cities. As deforestation became nearly complete, the ruling classes placed the remaining woodlands off-limits to commoners, designating them as royal forests and deer parks (hunting preserves). Laws excluding commoners from these lands were passed between 1000 and 1200 in Britain, and during the 1300s in France and Germany.

Between 1347 and 1352, a second wave of bubonic plague, apparently originating from central Asia, swept across Europe. During this **Black Death**, some 25 million people died, one-third of the prior population. Abandonment of farms caused some reforestation, although populations in most areas had already recovered to pre-plague levels by 1500. By 1600, only 5–15% of west-central Europe still remained forested, but forests still covered large areas of eastern Europe from Poland to Russia. By that time, these remaining eastern woodlands were being heavily cut to supply lumber for the flourishing economies in western Europe.

The quality of population estimates during the 1500s and 1600s improved in eastern Europe, with parish registers and tax counts in modern-day areas of Germany, Belgium, and the Czech Republic; hearth

FIGURE 4-15 Windmill technology was adopted in the Netherlands during the Middle Ages. *[age fotostock/SuperStock.]*

counts in the Balkans; and poll taxes in European Russia. Populations began to explode during this time, and even the 8–10 million deaths during the Thirty Years' War (1618–1648) represented only a brief dip in an upward trend.

Agricultural technologies also began to improve during this period. As early as the 1200s, boggy areas in northern Italy (Tuscany) had been drained, and the excess water was used to irrigate drier land. In the Netherlands, dykes were constructed to reclaim land from the sea for agriculture, and windmills were used to pump and move the drained water (Figure 4-15).

By the 1300s and 1400s, high population densities in the Netherlands forced farmers to innovate ways of producing more food from the land. The fallow period was shortened or eliminated by several new methods: spreading livestock manure; planting legume crops (peas and clover) to add nitrogen to depleted soil; rotating turnip and clover crops for use as winter fodder for animals; and using lime-rich sediment (marl) to sweeten the soil (reducing its acidity). These improved practices and more nutritious crops led to better animal husbandry: sheep with more fleece, cattle with more beef, and stronger horses for draft work. In addition, the **moldboard plow** arrived in the Netherlands from China, where it had been used for over 1,000 years (Figure 4-16). This more efficient plow, drawn by a team of horses or cattle, dug deep furrows in hard-to-work

FIGURE 4-16 Moldboard plow being used in Europe during the Middle Ages. *[The Pierpont Morgan Library/Art Resource, NY.]*

FIGURE 4-17 Deep furrows made by moldboard plows hundreds of years ago. *[Skyscan Photolibrary/Alamy.]*

soil, and turned the excavated earth upside-down to the sides. Traces of the deep furrows made by these plows are still visible in some fields (Figure 4-17).

All of these innovations greatly improved productivity and allowed farmers to make a living from smaller land holdings. By the 1600s, these new methods had spread to England, and then to other countries. In addition, imports of food and productive new crops like potatoes and corn from the Americas began to arrive in Europe by ship. By the 1800s, many countries obtained enough food from their land and from imports that farmers were able to let less-desirable holdings revert to forest.

Very late in pre-industrial times, burning of coal became widespread in some regions. By 1700, England was burning 3 million tons of coal per year, about a third of its energy use (but still a trivial amount from the standpoint of today's global CO_2 emissions). In a few northern European countries, peat was also burned for heating and cooking. By 1800, the population of Europe had increased to 180 million, due mainly to widespread improvements in sanitation and medicine. The Industrial Era technically began in Britain during the late 1700s, but emissions of Industrial Era CO_2 and CH_4 did not begin to rise sharply until the mid-1800s.

The history of deforestation in Europe reveals a slow but cumulatively striking shift. In the earliest phases many millennia ago, deforestation was viewed as a positive change: a means of opening up rich and fertile forest land to agriculture, and also a way to reduce habitat for unwelcome "wild beasts." Lions and leopards persisted in Greece and the Balkans until 2,000 years ago, and wolves and bears lived in many other regions until more recent times.

But by the Roman era, this early phase of deforestation appears to have eliminated half of the forests of Europe, leaving ever-smaller woodland areas mostly on steep and remote hillsides. By this time, the environmental damage caused by cutting hillside forests (erosion and silting of rivers) became more obvious, and deforestation began to be

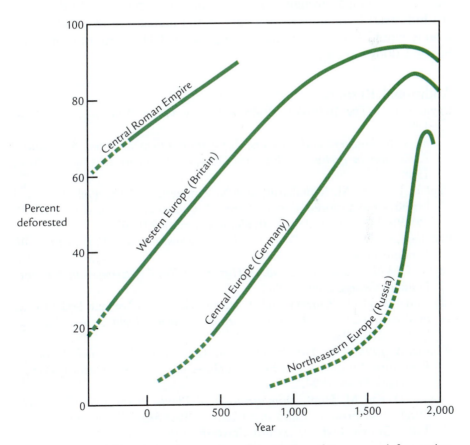

FIGURE 4-18 Highly schematic overview of the timing of European deforestation, from southwest to northeast.

viewed more as a problem than as a benefit. And by the time nobles claimed the remaining fragments of woodland for their own, the early view of forests as an unlimited resource had been long forgotten.

The timing of this shift in attitude varied from country to country, as deforestation followed somewhat different local trajectories. The classical civilizations of the Mediterranean region moved along the deforestation trajectory early in the historical era, followed by countries in western and central Europe roughly a millennium ago, and finally by those in far northeastern Europe in more recent centuries (Figure 4-18).

Summary
Compounded over thousands of years, this long history of deforestation in Europe must have contributed to the CO_2 rise that began 7,000 years ago and continued through most of the pre-industrial era into the modern Industrial Era. Similarly, the spread and increase in livestock populations would have contributed to the rise in CH_4 concentrations that began 5,000 years ago.

Additional Resources
Davis, S. J. M. *The Archeology of Animals*. New Haven: Yale University Press, 1995.

Gregg, S. A. *Foragers and Farmers: Population Interaction and Agricultural Expansion in Prehistoric Europe*. Chicago: University of Chicago Press, 1988.

Kaplan, J. O., K. M. Krumhardt, and N. Zimmerman. "The Prehistoric and Preindustrial Deforestation of Europe." *Quaternary Science Reviews* 28 (2009): 3016–3034. doi:10.1016/j.quascirev.2009.09.028.

Rackham, O. *Ancient Woodland: Its History, Vegetation and Uses in England*. London: Edwin Arnold, 1980.

Taylor, C. *Village and Farmstead: A History of Rural Settlement in England*. London: George Phillip, 1983.

van Andel, T. H., E. Zangger, and A. Demitrack. "Land Use and Soil Erosion in Prehistorical and Historical Greece." *Journal Field Archaeology* 17 (1990): 379–396.

Whittle, A. 1994. "The First Farmers," in *The Oxford Illustrated History of Prehistoric Europe*, edited by B. Cunliffe, pp. 136–166. Oxford: Oxford University Press, 1994.

Zohary, D. and M. Hopf. *Domestication of Plants in the Old World: The Origin and Spread of Cultivated Plants in West Asia, Europe, and the Nile Valley*, 3rd ed. Oxford: Oxford University Press, 2001.

CHAPTER 5

CHINA AND SOUTHERN ASIA

Because of its vast size, Asia spans many climate zones and vegetation types (Figure 5-1). Much of the continent was unfavorable to early agriculture: the tundra that borders the margin of the Arctic Ocean; the arctic forest (composed of larch trees; also called the *taiga*) that covers much of the rest of Siberia and the far north; the vast region of semi-arid grassland steppes across south-central Asia; and the deserts in parts of far southwest Asia. Areas more favorable for agriculture included the temperate deciduous forests along the eastern Pacific coast, and the evergreen **rainforests** and seasonally drier forests in the eastern tropics. The two most populous countries in today's world, China and India, already had very large populations by the start of historical times, partly because of favorable climates and partly because they had become centers of agricultural innovation early in the Neolithic Revolution (Table 5-1).

The history of agriculture in Asia encompasses a range of regional trends. Millet and soybeans were domesticated in northern China, along with pigs, whereas the domestication of irrigated rice occurred in south-central China. Beans, small forms of millet, and sesame were domesticated in India, as well as water buffalo and humped **Zebu cattle**. Several crops that had been domesticated in the semi-arid Fertile Crescent region to the west were later adapted, along with sheep and goats, by people in what is now northeast Pakistan and northwest India.

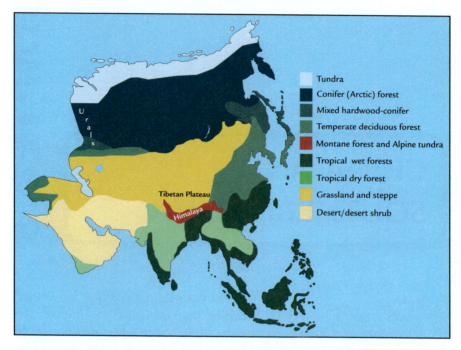

FIGURE 5-1 Major vegetation and mountains in Asia. *[Adapted from R. L. Smith and T. M. Smith,* Elements of Ecology *(Menlo Park: Benjamin/Cummings, 1998).]*

The Early Prehistorical Record

DNA evidence reveals two large-scale migrations of people who had left northeast Africa and entered arid regions of southwestern Asia (modern-day Iran) nearly 50,000 years ago. From this area, people gradually made their way into eastern Asia along two routes separated by the imposing Tibetan Plateau and Himalaya (Figure 5-1). One group of darker-skinned people took a southern (near-coastal) route nearly 40,000 years ago, moving through what is now Pakistan and India, and eventually into far southeastern Asia (modern-day Thailand, Indonesia, the Philippines, and southern China). A second, lighter-skinned group took a northern (mid-continent) route, heading east across the southern Siberian steppes between 40,000 and 30,000 years ago and then south into China, where they encountered people coming north from the southern migration route.

Early land use in the most heavily populated parts of Asia is hard to estimate, in part because subsequent activities of dense populations have overprinted much of the evidence. Nevertheless, China has

TABLE 5-1 Crop and Livestock Domestication in Southern/Eastern Asia

Crops	Time Domesticated	Region
Foxtail millet	By 8,000 years ago	Northern China
Common millet	By 8,000 years ago	Northern China
Rice	9,000–6,000 years ago	Central China
Soybeans	By 4,500 years ago	Northern China
Irrigated rice	8,000–6,000 years ago	South-Central China
Taro	By 4,500 years ago	Southern China
Apples	By 4,000 years ago	China
Pears	By 4,000 years ago	China
Plums	By 4,000 years ago	China
Mung beans	5,000–4,000 years ago	Western India
Urd beans	By 4,500 years ago	Western India
Little millet	By 5,000 years ago	Western India
Sesame	By 4,500 years ago	Western India
Buckwheat	By 5,000 years ago	Northeast India

Livestock	Time Domesticated	Region
Pigs	By 9,500 years ago	Northern China
Chickens	By 4,500 years ago	China, India
Horses	By 6,000 years ago	Central Asia
Cattle (Zebu)	9,000–7,000 years ago	Western India
Water buffalo	By 5,000 years ago	Western India

accumulated a great trove of archeological and historical information that can help us reconstruct the history of its population and its past land use. In contrast, India and other regions in southern and southeastern Asia have relatively little historical data and have not yet been studied as extensively.

Dryland Agriculture and Livestock in China

Evidence of agriculture in northern China dates to 9,500 years ago, and an even earlier origin is possible. Early farmers in that region used polished stone axes similar to those in the Fertile Crescent and Europe

FIGURE 5-2 The grain millet, domesticated in China. *[Robert L. Smith.]*

(recall Chapter 4). The first domesticated crops included two forms of millet (Figure 5-2) as well as dryland rice that did not require irrigation. Millet farming was centered along the Yellow (Huang He) River in the northern part of China, with rice farming more common near the Yangtze River to the south. At that time, the wet summer monsoon circulation in China was stronger than today and provided ample rainfall for growing crops (recall Chapter 2, Figure 2-5).

By 8,500 to 7,000 years ago, early Chinese farmers had domesticated pigs, chickens, and dogs (Figure 5-3) and were using polished stone tools comparable to those appearing in the Fertile Crescent. Holes for planting seeds were made by poking dibble sticks into the soil. Farmers used stone plows to break up the soil, and dug into it with spades made of bone (Figure 5-4). One difference from the Fertile Crescent is that sickles for harvesting grain crops had not yet been brought to or invented in China. Instead, knives may have been used to harvest grains, or plants may simply have been pulled up by their roots. By 6,000 to 5,000 years ago, northern Chinese farmers had also domesticated soybeans. Later, wheat, barley, and other crops arrived from the Fertile Crescent, along with cattle and some sheep and goats. Coastal regions generally relied more heavily on maritime food sources (fish and shellfish).

FIGURE 5-3 Pigs, domesticated in China.
[Nelson Ching/Bloomberg via Getty Images.]

FIGURE 5-4 Early farming tools in China: (A) stone plow with a wooden base; (B) 6,500-year-old bone spade adapted from the scapula of a water buffalo. *[A courtesy of Dr. Ling Qin of Peking University and Professor Xin-min Xu of Zhejiang Institute of Archaeology and Cultural Relics. B courtesy of Professor Guo-ping Sun, Zhejiang Institute of Archaeology and Cultural Relics.]*

One result of these early farming activities can be seen in a comprehensive summary of more than 11,000 archeological sites across much of east-central China (Figure 5-5). Relatively few sites have been found prior to 8,000 years ago, suggesting that only small populations existed in the region at that time. Between 8,000 and 7,000 years ago, the number of sites was still small but had begun to increase in east-central China along and south of the Yellow River. Between 7,000 and 5,000 years ago, sites spread north and west of the Yellow River and south into the Yangtze River valley, as well as into coastal regions. By 5,000–4,000 years ago, the density of sites had increased greatly, with evidence of more than fifty walled cities in the archeological record from this time period.

If we assume that the density of archeological sites over thousand-year intervals is very roughly proportional to the population of China, then the forty-fold increase in sites from the interval spanning 8,000 to 7,000 years ago to the interval spanning 5,000 to 4,000 years ago implies an enormous increase in population. By this (crude) index, population may have doubled five to six times in 3,000 years, for an average doubling time of approximately 600 years. This very rapid increase appears to be linked to the spread of early agriculture during this time period.

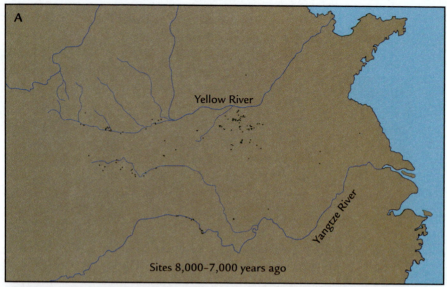

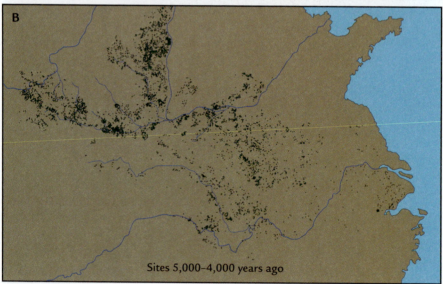

FIGURE 5-5 Increased number of archeological sites in central China 8,000 to 7,000 and 5,000 to 4,000 years ago. *[Adapted from X. Li, J. Dodson, J. Zhou, X. Y. Zhou, "Increases of Population and Expansion of Rice Agriculture in Asia, and Methane Emissions Since 5000 BP,"* Quaternary International *202 (2009): 41–50. doi:10.1016/j .quaint.2008.02.009.]*

At this time, most people in China were farmers, and this rapid population growth occurred mainly in forested east-central regions. As a result, the amount of forest cleared to make agriculture possible must have increased greatly between 8,000 and 4,000 years ago, which, similar to changes in Europe, brackets the time near 7,000 years ago when the atmospheric CO_2 trend reversed its downward trajectory and began to rise (recall Chapter 3, Figure 3-7).

Dryland Agriculture and Livestock on the Indian Subcontinent

Western India and eastern Pakistan also have a long history of dryland agriculture. The earliest cultures in this region date to 7,000 years ago, and the Harappan civilization (one of the largest Bronze Age cultures) had developed by 4,600 to 3,600 years ago (Figure 5-6). Because this region is semi-arid, urbanized centers formed near reliable sources of water provided by tributaries of the Indus River flowing south and southwest from the western Himalaya. These centers became places of organized food storage.

By 7,000 years ago, the Fertile Crescent package of crops (wheat, barley, peas, lentils, and chickpeas) had spread eastward into what is now Pakistan and northwest India (Figure 5-7), and by the time of the Harappan civilization, these crops were widely cultivated, alongside domesticated goats and sheep. Sorghum and pearl millet that had first been domesticated in Africa also appeared later.

Local domestication of humped Zebu cattle occurred between 9,000 and 7,000 years ago (Figure 5-8), and water buffalo were domesticated by 4,500 years ago. Locally domesticated crops included a small local variety of millet and dry Indica rice by 5,000 years ago, as well as sesame, eggplant, and mung and urd beans by 4,500 years ago (Figure 5-9).

As the summer monsoon circulation gradually weakened over southern Asia, rainfall in the western Himalaya decreased. Some archeological evidence suggests that a reduction in flow of the Indus River because of changes in the local monsoon climate may have caused the collapse of the Harappan civilization after 4,000 years ago, but other evidence points to shifts in river channels that left cities stranded far from sources of water (see Figure 5-6). One major channel of the Indus River may have been captured by the Ganges River and rerouted to the east. In any case, the center of population in India moved east toward the Ganges-Brahmaputra river valley. Subsequently, mung and

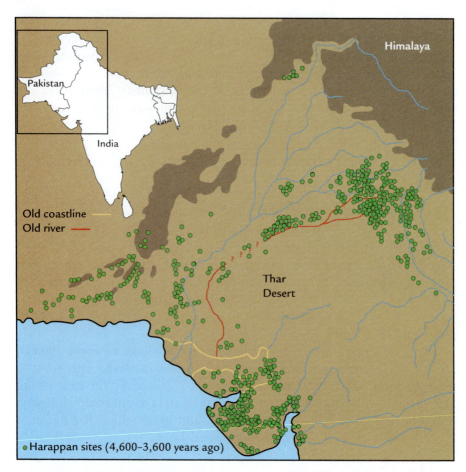

FIGURE 5-6 Harappan civilization sites near the Indus River in western India. *[Adapted from M. Medella and D. Q. Fuller, "Paleoecology and the Harappan Civilisation of South Asia: A Reconsideration,"* Quaternary Science Review *25 (2006): 1283–1301. doi:10.1016/j .quascirev.2005.10.012.]*

urd beans spread east into the Ganges River area of eastern India, and then on to Thailand and other areas of southeast Asia.

Spread of Livestock in Southern Asia

Archeobotanist Dorian Fuller and colleagues synthesized archeological data and mapped the spread of domesticated livestock (mostly cattle, with some sheep and goats) across southern Asia (Figure 5-10). Sites from between 7,000 and 5,000 years ago have been found with livestock

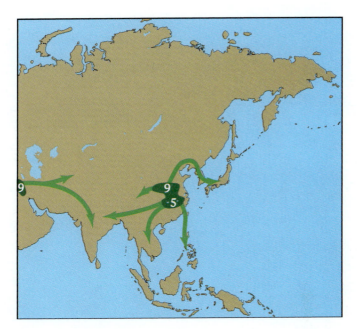

FIGURE 5-7 Spread of agriculture into ancient India from the Fertile Crescent and from China. Numbers show thousands of years ago. *[Adapted from P. Bellwood, First Farmers: The Origins of Agriculture (Oxford: Blackwell, 2004); and from M. D. Purugganan and D. Q. Fuller, "The Nature of Selection During Plant Domestication," Nature 457 (2009): 843–848. doi:10.1038/nature07895.]*

remains mainly in very arid regions: Arabia, the high Iranian Plateau, western India, and far western China. In these dry regions, pastoral people tending herds of livestock lived a nomadic life, moving constantly in search of fresh grass and water. Population densities there never grew very high, because the lack of rainfall limited the capacity of the land to sustain large herds and human populations.

Between 5,000 and 3,000 years ago, livestock spread across most of India, China, and the rest of southeast Asia, and then into the remaining areas between 3,000 and 1,000 years ago. Sheep, goats, and cattle (nonhumped) that had been domesticated in southwest Asia first appeared in India after 5,000 years ago, and reached northern China by 4,000 to 3,000 years ago. During the same interval, Zebu cattle domesticated in eastern Pakistan spread southeastward across India (Figure 5-11). Archeological data in southeast Asia, between India and China, are scarce but also indicate the appearance of livestock between 5,000 and 1,000 years ago.

FIGURE 5-8 Zebu cattle domesticated in India. *[blickwinkel/Alamy.]*

The expansion of livestock after 5,000 years ago occurred mainly in well-watered regions where reliable rainfall was delivered by the summer monsoon. The rapid growth in farming populations in these regions was likely accompanied by comparable increases in livestock because of the greater capacity of rain-fed pastures to sustain large

numbers of animals. The resulting increase in livestock across all of southern and eastern Asia coincides with the reversal in atmospheric CH_4 concentration 5,000 years ago and its subsequent upward trend (recall Chapter 2, Figure 2-8). Livestock in this period were a growing methane source, adding to the contribution from the spread of Fertile Crescent livestock into Europe

FIGURE 5-9 Mung beans, domesticated in India. *[Robert L. Smith.]*

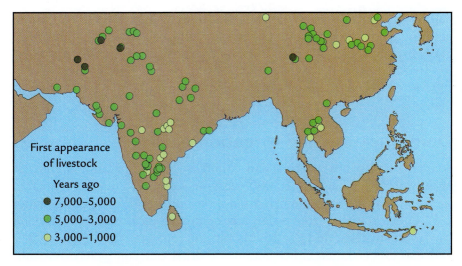

FIGURE 5-10 Spread of domesticated livestock across India, southeast Asia, and China. *[Adapted from D. Q. Fuller et al., "The Contribution of Rice Agriculture and Livestock to Prehistoric Methane Levels: An Archeological Assessment,"* The Holocene *25 (2011) 743–759. doi:10177/0959683611398052.]*

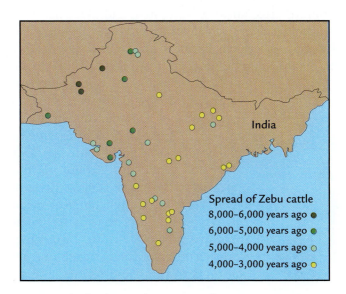

FIGURE 5-11 Spread of Zebu cattle across India. *[Adapted from S. Chen et al., "Zebu Cattle Are an Exclusive Legacy of the South Asia Neolithic,"* Molecular Biology and Evolution *27 (2010): 1–6. doi:10.1093/molbev/msp213.]*

before and during the same interval (recall Chapter 4). In addition, because all of these regions in southern and southeast Asia had originally been forested, clearing of the land to create pastures and hayfields for livestock added to CO_2 emissions during the time when atmospheric CO_2 concentrations were also rising rapidly (see Chapter 3, Figure 3-7).

Domestication and Irrigation of Rice

Southern and southeastern Asian countries shared a common climatic backdrop during the last 10,000 years—a slow but cumulatively large weakening of the wet summer Asian monsoon (recall Chapter 2, Figure 2-5). Pollen remains from Asian lakes show tropical wet forests giving way to dry forests, and dry forests turning into grasslands, while lakes and natural wetlands either shrank or dried out completely. As summarized in Chapter 2, this change in climate caused a loss of methane-emitting tropical wetlands and contributed to a steady natural decrease in atmospheric CH_4 concentrations after 10,000 years ago.

In China, rice had been grown for millennia as a dryland crop planted in well-watered areas and completely dependent on natural (monsoonal) rainfall (Figure 5-12). During that time (10,000 to 5,000 years ago), rice was not a major part of the Chinese diet. Methane emissions from dry rice farming were small and likely had no detectable effect on atmospheric CH_4 concentrations. Later, as the summer monsoon rains began to weaken, farmers in China responded with one of the most momentous agricultural innovations in history by domesticat-

ing irrigated rice, now one of the world's three major grain crops (along with maize and wheat). Today, almost one-quarter of the cultivated land in China is planted in irrigated rice.

The earliest known signs of domestication of the local Japonica variety of irrigated rice appeared in the lower Yangtze River basin some 8,000 years ago. Archeological evidence from this region indicates that early human populations had previously relied on a variety of naturally available foods, such as acorns and chestnuts from upland forests, and a variety of

FIGURE 5-12 Irrigated rice, domesticated in China. *[Robert L. Smith.]*

foods from low-lying and water-covered areas, including water chest-nuts, foxtail nuts, and fish (mostly carp). These hunter-gatherers also hunted deer, buffalo, and wild boar.

The wild precursor of domesticated rice dispersed its seeds the same way that wheat and many other grains do—by shattering (bursting off the plant stalks). The early stages of the process of domesticating ir-rigated rice involved artificial selection for non-shattering forms, as well as for stalks that grew in a more upright form. Irrigated rice is a high-productivity crop that yields about four times as much food per hectare farmed as dryland crops do, but it is also very labor-intensive. To cul-tivate this crop, farmers used fire to clear natural vegetation, and then leveled the soil in the paddy field areas. Young rice plants were planted in small nursery beds and then hand-transplanted to already flooded paddy fields. Water was moved by gravity—in ditches or along wooden sluiceways—from streams and rivers to the rice paddies. Where neces-sary, bucket-lifting systems moved water uphill. In some regions, wells were also used as water sources.

By 6,500 years ago, widespread adoption of irrigated rice in the middle reaches of the Yangtze River valley had begun to replace the prior reliance on hunting and gathering, and irrigated rice began to spread across China. At this time, most paddy fields were small ponds just a few meters on a side (Figure 5-13). Yangtze pollen records of this

FIGURE 5-13 Early rice-paddy ponds in the Chuodun fields region. *[Courtesy of Professor Jin-long Ding of Suzhou Museum, reproduced from Fuller and Qin, 2009 (see Additional Resources for full citation).]*

period show a decrease in pollen from elm trees, which lived in natural wetland forests. Climate change does not appear to explain this shift; anthropogenic burning to clear areas for rice agriculture is the likely cause. The shift to rice agriculture is also shown in the archeological record by large increases in **phytoliths**, small fragments made of silica that form part of many plants (mostly grasses), and survive the decomposition that destroys the rest of the plants after they die. The rapid increase of phytoliths from wet rice helps us trace the spread of irrigated agriculture in China and other parts of Asia.

Between 5,500 and 4,500 years ago, rice irrigation began a major regional expansion that would eventually cover all of southern Asia (Figure 5-14). At this time, farmers were also switching from small rice

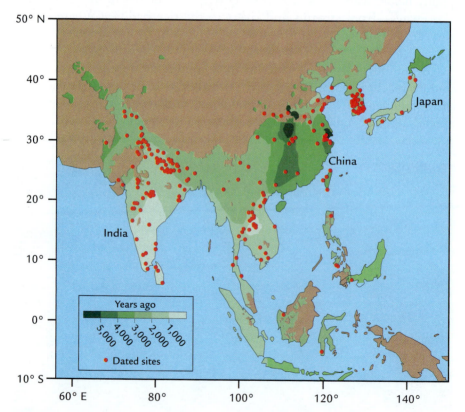

FIGURE 5-14 Spread of irrigated rice in southern and southeastern Asia.
[Adapted from D. Q. Fuller et al., "The Contribution of Rice Agriculture and Livestock to Prehistoric Methane Levels: An Archeological Assessment," The Holocene 25 (2011) 743–759. doi:10177/0959683611398052.]

ponds to larger paddy fields. Rice-based agriculture spread southward, accompanied by the Han culture and language from north-central China. Previous cultures and languages across southern China were either fragmented or disappeared completely.

Because rice paddies are a natural source of methane (recall Chapter 2), the spread of irrigated rice across a large area of Asia that began between 5,000 and 4,000 years ago must have caused a significant increase in CH_4 emissions. This increase began just at the time the atmospheric methane concentration stopped falling and began its wrong-way upward trend (see Chapter 2, Figure 2-8).

By 5,000 to 4,000 years ago, major cultural advances were occurring across much of China, and fortified towns surrounded by defensive walls appeared. The Bronze Age began in China nearly 4,500 years ago, and stronger and more efficient axes, plows, and other tools replaced stone predecessors. As Chinese society became more advanced, the first dynasty in China—the Xia—began 3,500 years ago.

One intensively studied site in Yunnan Province, far to the southwest of the more populous eastern China, indicates the breadth of changes underway during this period. Low-lying terrain near lakes and rivers shows initially minor forest clearance near 7,000 years ago, followed by much larger-scale clearance by 4,900 years ago. By this time, pollen evidence documents a change from nature-dominated tree pollen to clearance vegetation like grasses, *Artemisia*, and *Plantago*. The evidence even shows substantial clearance on higher slopes. By 3,000 to 2,500 years ago, rice irrigation was underway in the lower terrain of what is today Yunnan Province.

Irrigated rice spread from China to Korea and Japan, along with Bronze Age tools (see Figure 5-14). By 2,000 years ago, it had spread to India and across much of the rest of the southeastern Asian mainland (modern-day Thailand, Cambodia, Vietnam, and Malaysia) as well as offshore islands (modern-day Taiwan, Indonesia, and the Philippines). Archeological data from southeast Asia show early settlement of low-lying river valleys and floodplains that would have been favorable for growing rice, and pollen records indicate reduced amounts of natural swamp vegetation. In India, water buffalo used by farmers to till the land and trample dead rice stubble into the paddy soil were domesticated by 4,500 years ago and spread to other parts of southeast Asia by 3,000 years ago.

To supplement rice as a source of food, by 4,000 years ago farmers had developed the technique of grafting fruit trees (such as apples, pears, and plums) and planting them in terrain that was not suitable for rice paddies. Flooded paddy fields were also used as fish ponds (mostly for carp) after 2,500 years ago.

The Historical Record

In China and India, the historical era with its written records began more than 3,000 years ago. The early history of India is sparse, but some records are available from the urban-centered Mauryan civilization between 2,400 and 2,200 years ago, and from the Gupta era between 1,800 and 1,700 years ago. Both cultures were short-lived, and neither one led to sustained dynasties. The Mauryans grew dryland crops like wheat and barley that had arrived from the Fertile Crescent, along with an indigenous form of dry Indica rice that had been gathered in that region for thousands of years. Some population counts are available from the Gupta era based on village, monastery, and army registers, and these data can be scaled-up to estimate national totals. But then the record from India goes blank until the Moghul era in the 1600s.

Some investigators have claimed an independent origin for irrigated rice in western India, followed by a later spread eastward into the Ganges River valley, but recent archeological and archeobotanical evidence does not show local development of nonshattering forms of rice. Instead, the evidence indicates that irrigated rice moved westward from its origin in China to the Ganges River basin after 3,000 years ago, and then south across central India between 3,000 and 1,000 years ago (see Figure 5-14).

In contrast to India, China has a wealth of historical information on population, land use, and technology. The first written records appeared by 3,300 years ago, soon after the first (Xia) dynasty, and a short time before the Zhou dynasty that unified China between 3,100 and 2,200 years ago. Construction of the Grand Canal linking northern and southern China began soon after 2,500 years ago, and China's first emperor ruled during the Qin dynasty between 2,300 and 2,200 years ago. The first reliable census counts from the Han dynasty, just after 2,000 years ago, recorded 11.8 million households. With the standard assumption of five people per household, this number translates to almost 60 million people at that time.

Many naturally forested areas in eastern China had already become densely populated by 2,000 years ago. As the land became more crowded, farmers were forced to invent new ways to produce more food from smaller plots of land. Initially, cultivated fields had been fertilized mainly by mixing vegetable debris into the soil, but the increased use of live-stock allowed people to spread animal manure, along with "night soil" from the growing human population. The use of manure reduced the time during which plots had to lie fallow between plantings in order to regain their fertility. Farmers also learned to rotate their crops to im-prove soil quality. By 2,000 years ago, farmers were beginning to plant more than one crop per year, usually wheat in the cold season and rice in the warm season. Iron moldboard plows, which were more effective at turning soil than their predecessors, came into use about 2,000 years ago, long before their adoption in Europe around the year 1500.

From 2,000 to 1,000 years ago, irrigated rice farming spread to all remaining regions in southern and southeastern Asia where rice could be grown (see Figure 5-14). Also, within each region, more and more suitable land was converted to growing this nutritious crop. These two trends resulted in ever-more irrigated land and caused further increases in methane emissions during the interval when CH_4 concentrations in the atmosphere were rising.

After 2,000 years ago, farmers also began to build terraces that enclosed tiny rice paddies, forming small stair steps up steep hillsides (Figure 5-15). These precarious structures added little arable land and required an enormous amount of effort to build. It seems likely that all of the more easily flattened and farmed land in the large valleys below must already have been put into use as rice paddies, forcing the still-growing farming population to make use of marginal land.

The ongoing spread of livestock across southeastern Asia and the subsequent growth in the size of the herds also added to the anthro-pogenic methane emissions (which, on a global basis, are larger for livestock than for rice paddies today). The spread of crops would also have led to more extensive an-nual burning to clear leftover stubble from the previous year's crops, in order to keep cultivated fields free of grass, weeds, and pests, and to fertilize the soil to boost the next year's crop yield. Vegetation, buried in the deeper layers of piles of crop debris,

FIGURE 5-15 Asian farmers began to build small rice terraces on steep hillsides during the last 2,000 to 1,000 years. [Skip Nall/Getty Images.]

that burned in the partial absence of oxygen would have contributed yet more methane to the atmosphere.

Historical records from China provide reasonably reliable information on land-use change through the last 2,000 years. In 1937, economist John Buck compiled records of the amount of land that had been actively cultivated since the Han dynasty, expressed in units of **mou**, with one mou equivalent to about 600 square meters, or just over one hectare (2.2 acres). His compilations, reevaluated in 1989 by Kang Chao to adjust for changes in the size of the traditional mou unit through time, show a gradual increase in the total area of cultivated cropland in China during the last two millennia (Figure 5-16). But historically available census counts for this interval show an even larger increase in population. Because much of the arable land was already in use, the growing farming population had to make do with less cultivated land per person.

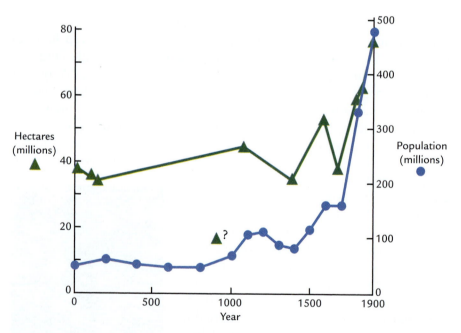

FIGURE 5-16 Increases in cultivated land and population of China between the year 5 and the 1800s. (The anomalously low clearance value at the year 976 followed shortly after a major war that left much of the land temporarily unusable.) [*Adapted from J. L. Buck,* Land Utilization in China *(Shanghai: Commercial Press, 1937); and K. Chao,* Man and Land in Chinese History: An Economic Analysis *(Stanford: Stanford University Press, 1986).]*

Because Chinese farmers followed the cultural practice of dividing their land among their surviving sons, their land holdings were broken up into ever-smaller pieces. Sustained over thousands of years, this ancient practice helps to explain today's typical Chinese "garden agriculture": small plots farmed by the manual labor of rural people, rather than large plots farmed by mechanization.

Long-term records of rice irrigation are also available from a small region in the lower Yangtze River basin that has long been one of the most productive rice-growing areas in China. In this region, all of the arable land was planted in rice by the years 1400 to 1600, yet populations continued to rise after this time (Figure 5-17). As population densities reached very high levels, the number of livestock kept per family began to decline. With ever-shrinking land holdings, farmers were gradually forced to choose between keeping livestock (which requires large amounts of land for summer pasture and winter fodder) and growing rice (which yields much more nutrition from the same amount of land). Gradually, over time, the farmers chose to keep fewer livestock and use them mainly as draft animals and sources of manure rather than for food. In a reversion to ancient practices, many farmers also reduced their dependency on livestock and in many cases pulled their own plows.

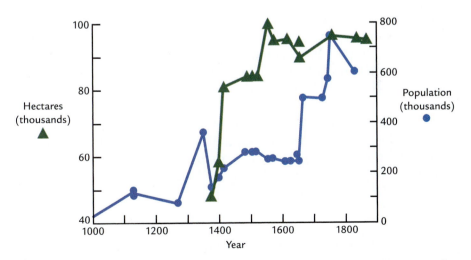

FIGURE 5-17 Increases in rice-paddy areas farmed and human population in the Tai Lake region of the Yangtze River valley between the year 1000 and the 1800s. *[Adapted from E. C. Ellis and S. M. Wang, "Sustainable Traditional Agriculture in the Tai Lake Region of China,"* Agriculture Ecosystems and Environment *61 (1997): 177–193.]*

The water buffalo and oxen that still remained were shared among neighbors in villages and fed on stubble from cultivated fields.

As this more intensive use of the land developed, farmers created new ways to grow more food per acre: greater application of manure as fertilizer (from vegetation, livestock, and humans), increased human labor to combat weeds and pests, and wider use of the iron moldboard plow (including even some use of steel plows in the Tang dynasty during the 900s). A new, faster-growing variety of rice (Champa) from Vietnam that appeared after the year 800 allowed farmers in southern China to grow two crops of rice during the warm season rather than one, and farmers farther to the north were able to harvest rice during their short summers. Much of the additional land brought under cultivation after about 1400 came from reclamation of swampy coastal areas and attempts to farm substandard hill slopes and sandy soils.

All of these trends suggest very intensive deforestation in densely populated eastern China after 2,000 years ago. In response to this development, coal had come into use as a source of home heating by the year 400, if not earlier, because of its ready accessibility from beds in shallow open-pit mines. In the Kaifeng region of east-central China, a local observer noted during the 1100s that ". . . no dwelling burns wood any more." By that time, the Sung dynasty had developed a vigorous iron industry that relied heavily on wood for making charcoal for furnaces. At various times during the last 2,000 years, emperors issued edicts to preserve remaining forest remnants from peasant use.

Aside from China, historical land-use data are scarce for southeast Asia. Even reliable census counts did not begin until relatively recent centuries in the rest of this region. The earliest household counts in Japan date to the 800s and in Korea to the 1000s. The first reliable population estimates in southwest Asia are even more recent, with counts of households (or males) beginning in Turkey, Syria, Lebanon, and Iraq during the 1500s, but not until the 1900s in Iran and Arabia. (Sumeria [modern Iraq] is known to have been populous prior to 2,000 years ago, but it probably never amounted to more than 15% of the total population of Asia.)

Summary

Archeological records across Asia show that the spread of agriculture coincided with the anomalous upward CO_2 and CH_4 trends revealed by ice cores. China originally had more than 4 million square kilometers of forests in its eastern half, but clearance of forests (mainly for dryland agriculture) grew rapidly between 8,000 and 4,000 years ago as populations increased in central China (see Figure 5-5). This rapid increase in archeological sites brackets the reversal of the CO_2 trend nearly 7,000 years ago. By 2,000 years ago, China was densely populated with 50 to 60 million people, and most of its northeastern forests had probably been cut. Since 2,000 years ago, average landholdings per person have fallen by a factor of four as farmers have continually subdivided their land among multiple male heirs. Major deforestation probably occurred somewhat later in southern China and other parts of southeast Asia, but forest clearance in much of China appears to have been nearly complete by the years 1400 to 1600.

The spread of methane-emitting rice irrigation began nearly 5,000 years ago (see Figure 5-14), the time of the reversal of the CH_4 trend. The subsequent spread of rice agriculture across all of southern Asia matches the continuing rise in CH_4 concentrations, as does the spread of livestock across Asia and Africa (see Figures 5-10 and 5-11). As populations across Asia continued to increase, emissions from other CH_4 sources like biomass burning and human waste also grew.

Additional Resources

Buck, J. L. *Land Utilization in China*. Shanghai: Commercial Press, 1937.

Chao, K. *Man and Land in Chinese History: An Economic Analysis*. Stanford: Stanford University Press, 1986.

Dearing, J. A., R. T. Jones, J. Shen, X. Yang, J. F. Boyle, G. C. Foster, D. S. Crook, and M. J. D. Elvin. "Using Multiple Archives to Understand Past and Present Climate-Human-Environment Interactions: The Lake Erhai Catchment, Yunnan Province, China." *Journal of Paleolimnology* 40 (2008): 3–31, doi:10.1007/s10933-007-9182-2.

Ellis, E. C., and S. M. Wang. "Sustainable Traditional Agriculture in the Tai Lake Region of China." *Agriculture Ecosystems & Environment* 61 (1997): 177–193.

Elvin, M. "Three Thousand Years of Unsustainable Growth: China's Environment from Archaic Times to the Present." *East Asian History* 6 (1993): 7–46.

Fairservice, W. A. Jr. *The Roots of Ancient India*. New York: Macmillan, 1971.

Flint, E. P., and J. F. Richards. "Historical Analysis of Changes in Land Use and Carbon Stock of Vegetation in South and Southeast Asia." *Canadian Journal of Forest Research* 21 (1991): 91–110.

Freese, B. *Coal: A Human History*. New York: Penguin Books, 2003.

Fuller, D. Q., and L. Qin. "Water Management and Labour in the Origins and Dispersal of Asian Rice." *World Archeology* 41 (2009): 88–111, doi:10.1080/00438240802668321.

Fuller, D. Q., J. Van Etten, K. Manning, C. Castillo, E. Kingwell-Banham, A. Weisskopf, L. Qin, Y.-I. Sato, and R. Hijmans. "The Contribution of Rice Agriculture and Livestock to Prehistoric Methane Levels: An

Archeological Assessment." *The Holocene* 25 (2011): 743–759, doi:10177/0959683611398052.

Harlan, J. R. *The Living Fields: Our Agricultural Heritage.* Cambridge: Cambridge University Press, 1995.

Li, X., J. Dodson, J. Zhou, and X. Y. Zhou. "Increases of Population and Expansion of Rice Agriculture in Asia, and Methane Emissions Since 5000 BP." *Quaternary International* 202 (2009): 41–50, doi:10.1016/j.quaint.2008.02.009.

Ruddiman, W. F., Z. Guo, X. Zhou, H. Wu, and Y. Yu. "Early Rice Farming and Anomalous Methane Trends." *Quaternary Science Reviews* 27 (2008): 1291–1295, doi:10.1016/j.quascirev.2008.03.007.

CHAPTER 6

THE AMERICAS

The third major center of agricultural innovation was the Americas—mainly modern-day Mexico, the Peruvian Andes, and the Amazon Basin. Familiar foods domesticated in those regions include corn (widely known as maize), squash, potatoes, sweet potatoes, several kinds of beans, manioc, tomatoes, and peanuts (Table 6-1). Archeological research has steadily pushed the origins of these domesticated crops back in time, some almost as far back as those in the Fertile Crescent and in northern China. Few animals in the Americas were suited by size or temperament for domestication except for llama and alpaca.

During the early 1900s, most archeologists thought that prior to European contact the Americas were sparsely populated by perhaps fifteen million people who had little impact on the environment and "lived lightly on the land." In recent decades, however, a different picture has emerged. Early American populations were much larger than was originally thought, but were nearly wiped out by diseases introduced by initial contact with Europeans and their livestock. This calamity is now estimated to have killed 85% to 90% of a pre-contact population of approximately forty to sixty million people.

Most early Americans were agriculturalists, and early farming required extensive clearing or modification of forested land. By the late 1400s, many parts of the Americas were not nature-dominated environments but were substantially transformed anthropogenic landscapes. Evidence for this new vision of agricultural America comes from accounts written by early European visitors and from archeological and ecological evidence.

TABLE 6-1 Crop and Livestock Domestication in the Americas

Crops	Time Domesticated	Region
Squash	By 10,000 years ago	Central America
Corn (Maize)	9,000–7,000 years ago	Central America
Beans (one type)	By 4,300 years ago	Central America
Tomato	By 4,300 years ago	Central America
Avocado	By 4,300 years ago	Central America
Acorn squash	9,000–8,000 years ago	Northern South America
Sweet potato	9,000–8,000 years ago	Northern South America
Common bean	9,000–8,000 years ago	Northern South America
Potato	By 8,000 years ago	Upper Peruvian Andes
Quinoa	By 5,000 years ago	Upper Peruvian Andes
Lima bean	By 10,000 years ago	Lower Andes
Chili peppers	By 6,000 years ago	Lower Andes
Peanuts	By 8,500 years ago	Lower Andes/Amazon
Manioc (cassava)	By 8,000 years ago	Lower Andes/Amazon
Fruit tree orchards	Starting 11,000 years ago	Lower Amazon Basin
Oil palm nuts	Starting 11,000 years ago	Lower Amazon Basin
Peach palms	By 2,300 years ago	Lower Amazon Basin
Sumpweed	4,500–4,000 years ago	Eastern North America
Goosefoot	4,500–4,000 years ago	Eastern North America
Maygrass	4,500–4,000 years ago	Eastern North America
Sunflower	4,500–4,000 years ago	Eastern North America

Livestock	Time Domesticated	Region
Llama	By 6,000 years ago	Peruvian Andes
Alpaca	By 6,000 years ago	Peruvian Andes

Because the Americas span latitudes from the Arctic all the way south to the sub-Antarctic and are fragmented by several mountain ranges, the variety of vegetation types is enormous (Figure 6-1). In North America, tundra borders Hudson Bay and the Arctic Ocean, conifer (spruce and fir) forests cover most of Canada, and deciduous hardwood forests dominate the eastern United States, giving way westward to prairie grasslands in the plains, and to deserts farther to the southwest. Both grasslands and deserts

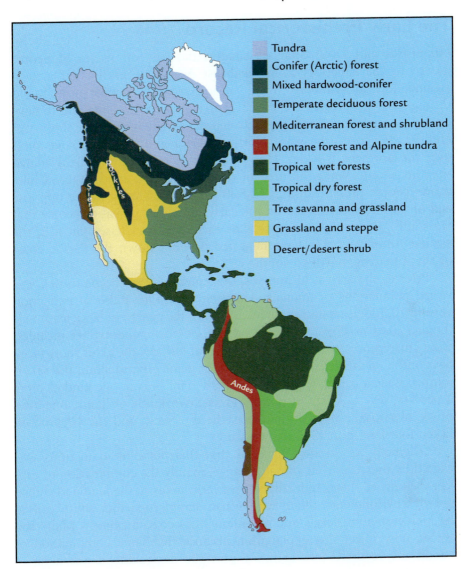

FIGURE 6-1 Major vegetation groups and mountains in the Americas. *[Adapted from R. L. Smith and T. M. Smith,* Elements of Ecology *(Menlo Park: Benjamin/Cummings, 1998).]*

extend south into parts of northern Mexico. Most of the isthmus of Central America is tropical rain forest, as is most of the vast Amazon Basin in central South America. Bordering these rain forests are larger areas of dry tropical forest, tree savanna, and grasslands, including the pampas of Argentina. Montane forests cover much of the Andes, with alpine tundra at high altitudes. Parts of the Pacific coast of South America are desert.

The Early Prehistorical Record

Linguistics and DNA analyses suggest that several waves of people moved to the Americas from northeast Asia. One group that arrived around 15,000 years ago included ancestors of people now found throughout the Americas. These early arrivals were once thought to have passed through an ice-free gap between the slowly melting ice sheet over Hudson Bay in northeastern North America and the smaller ice sheet over the northern Rockies. However, one of the earliest reliably dated sites (in Monte Verde, an area of coastal southern modern-day Chile) shows people present around 12,800 years ago, almost at the opposite end of the Americas. Based on the early timing of arrival of people that far south, along with evidence that their diet was based on fish, the earliest Americans may have migrated by boat down the Pacific coast of the Americas, north to south.

Prior to the arrival of humans, the Americas were home to rich and varied populations of animals. Then, an enormous pulse of extinction around 12,500 years ago eliminated 75% of the genera of large mammals (human-sized or larger). The list of species now gone extinct includes: wooly mammoths, mastodons, two kinds of buffalo, the giant short-faced bear and a spectacled bear, four kinds of giant ground sloth, the saber-toothed tiger, a camel, a cheetah, a yak, a giant beaver, one kind of tapir, a peccary, several kinds of horse and ass, two kinds of llama, one kind of ox, a moose, a deer, an antelope, the dire wolf, and the *Glyptodont* (a ten-foot-long armadillo-like creature).

Originally, researchers believed that this pulse of extinctions was caused by pressure from climate change, mainly the combination of rapidly melting ice sheets that still covered large areas of the land and chilled nearby environments, along with warm summers farther south caused by high insolation levels and greenhouse-gas concentrations. Many scientists now reject this explanation because this same combination of conditions occurred late in many previous deglaciations, yet none of those intervals show any sign of unusual clusters of extinctions (see Chapters 1–3).

This observation raises an obvious question: Why would the climatic effects of this most recent deglaciation be so completely different? Based on the lack of a convincing answer to this question, some scientists have suggested that humans, and their growing skills at hunting and killing mammals, were the key. This idea that humans caused the extinction of many large mammals is known as the **overkill hypothesis.** One criticism of this theory is that the pulse of extinction occurred at least a few thousand years after the earliest human arrivals. Possible responses to this criticism counter that human populations took time to build to the levels needed to completely eliminate so many mammal species, or that the introduction of some kind of new hunting technology soon after human arrival may have been crucial.

The new arrivals from Asia were hunter-gatherers (and fishers) lacking bronze or iron tools. Yet several American cultures had begun to work copper by the time of European entry, and two of them (Aztecs in Mexico and Incas in Peru) had become massive empires. Also, as noted above, American peoples had become among the world's greatest agriculturalists (see Table 6-1). But because of geographic isolation (created by barriers of mountains and deserts) combined with very different climate zones and natural vegetation, many agricultural innovations remained localized, and agricultural patterns differed greatly from region to region.

Mesoamerica (Mexico and Central America)

Mesoamerica spans a range of climatic and vegetation zones (Figure 6-2, and see Figure 6-1). The region most relevant to the spread of agriculture includes the semi-arid central Mexican highlands, with dry tropical forests and runoff from mountain rains and snows, and wet tropical forests

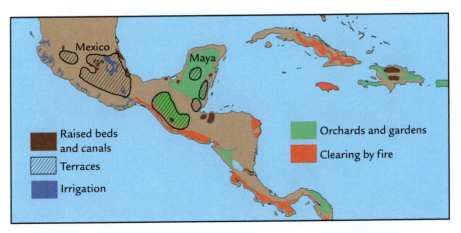

FIGURE 6-2 Centers of early agriculture in Central America. *[Adapted from map by C. C. Mann, W. Doolittle, and P. Dana, in C. C. Mann, 1491 (New York: Random House, 2006); courtesy of C. C. Mann.]*

FIGURE 6-3 Squash, domesticated in Central America. [Robert L. Smith.]

along the lowlands on the coasts of the Pacific Ocean and Gulf of Mexico. These regions were the center of domestication of an amazing variety of foods, including maize, squash, tomatoes, avocados, and several types of beans (Figures 6-3 and 6-4). The wild ancestors of tomatoes grew naturally in the areas now known as Peru and Ecuador as plants with tiny berries that were smaller than modern-day peas, but the slow domestication of these wild varieties to larger forms apparently occurred in what is now western Mexico.

The first agriculture in this region dates to 10,000 years ago or earlier, when several kinds of squash were first domesticated and grown in tropical lowlands. This date rivals the earliest agriculture in China and lags not far behind the earliest agriculture in the Fertile Crescent.

By 8,700 years ago, evidence of early domestication of maize—one of the three major modern-day grain crops in the world—appears in the Central Balsas River valley in southwest present-day Mexico. Stone grinding tools of this age still have traces of maize on them, and DNA analyses suggest that initial domestication occurred at about the same time.

The natural ancestor of maize was debated for decades, because nothing similar to it grows wild in the region today. Careful archeological work and DNA analyses have shown that maize

FIGURE 6-4 Beans, domesticated in Mexico. [Robert L. Smith.]

FIGURE 6-5 Gradual domestication of a wild form of grass called teosinte (left) through intermediate forms over many thousands of years led to the modern crop maize (right). [Courtesy of John Doebly.]

comes from a natural wild grass called **teosinte**, although the reason for its domestication remains something of a mystery. Wild wheat and rice have large grains that look like their domesticated successors, but teosinte has a dozen or so tiny "kernels" encased in a very hard outer casing (Figure 6-5). Why early farmers found teosinte worth their attention is difficult to understand. One possibility is that they were mainly interested in the sugar-rich stalks of the plant, not the kernels. Or perhaps they didn't actually pay much attention to teosinte at first, using it as a minor supplement to other sources of food. Instinctively, they may have chosen grains that had openings in the hard outer casings and that didn't exhibit the shattering behavior that made harvesting difficult. In any case, somehow, over many thousands of years, early American farmers ended up selecting for larger, juicier kernels.

As this process continued, maize remained a minor item in the diet of Mesoamericans for more than 4,000 years. Even though some degree of domestication had occurred by 8,700 years ago, the ears (cobs) were too small to be relied on as a major food source, and people continued to eat wild food sources along with crops like squash. Agriculture had not yet reached the point of providing the complete nutritional base that people needed.

By 4,300 years ago, ears of maize had become large enough for the crop to become sufficiently productive for greater, more widespread agricultural use. Combining maize with squash (domesticated earlier) and with beans, tomatoes, avocados, and melons (all domesticated later), people in what is now Central Mexico gradually developed a fully nutritional package of crops planted together in plots called **milpa**. Aside from its nutritional benefit, this combination of crops (particularly beans that can extract nitrogen from the air and store it in the soil) allowed land to remain fertile for relatively long periods of time. Eventually, maize and the other milpa crops spread south into the Amazon Basin and lower forests of the Andes and then north into

North America (Figure 6-6). Compared to Europe, however, the transition from a hunting and gathering life to a fully agricultural one was slow, in part because of the gradual pace of maize domestication, but also because wild animals suitable for domestication were not widely available for use in field labor.

In Mesoamerica, the success of milpa farming eventually made possible a sedentary life and the early emergence of several major civilizations based on agriculture (recall Figure 6-2). Between 3,500 and 2,500 years ago, the Olmec people on the Gulf Coast of Mexico formed one

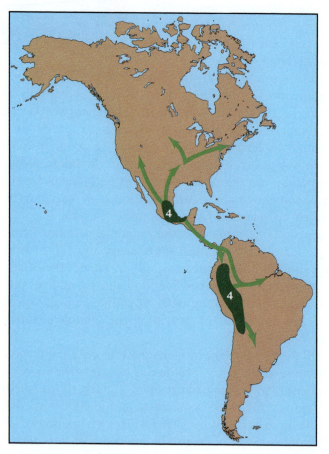

FIGURE 6-6 Dispersal of agriculture from centers in Mexico and the Andes. Numbers are thousands of years ago. [*Adapted from P. Bellwood,* First Farmers: The Origins of Agriculture *(Oxford: Blackwell, 2004); and from M. D. Purugganan and D. Q. Fuller, "The Nature of Selection During Plant Domestication,"* Nature *457 (2009): 843–848, doi:10.1038/nature07895.*]

of the first technologically sophisticated American societies, with towns and cities, colossal stone statues, and a cultural influence that gradually spread to many other regions. Stalks of maize featured prominently in Olmec art. Between 3,000 and 2,000 years ago, a series of regional empires developed, including the Toltecs in the central northern highlands, and the Zapotecs followed by the Mixtecs in the southern highlands and along the Pacific coast.

On the Yucatan Peninsula farther south, Mayan cultures emerged between 4,000 and 3,000 years ago, as people began cutting the coastal forests to cultivate maize. Mayan farmers planted two crops a year by terracing higher terrain to direct runoff from rains, and by constructing raised fields in waterlogged terrain separated by submerged ditches and canals that supplied vegetation for use as fertilizer. An increase in Mayan populations in the frequently warring city-states caused heavy deforestation between 3,000 and 1,000 years ago.

The archeologist Arlen Chase and colleagues recently studied the large Mayan city of Caracol in the modern-day country of Belize, using **LIDAR** (*Li*ght *D*etection *A*nd *R*anging), a technology that can peer through the jungle canopies and reveal structures lying below them (Figure 6-7). Previously, twenty-five years of painstaking ground-based archeological fieldwork at this site had identified several major buildings, connecting roads, and terraced areas. In just four days, this new LIDAR survey thoroughly mapped an area more than seven times larger, down to the scale of features such as small buildings, causeways, and very extensive agricultural terraces. Future use of this technique may bring many more such revelations about other tropical rain forest areas. Based on this LIDAR survey, we know that at its peak, between the years 500 and 900, Caracol had a population of more than 100,000 people, with all the land for miles around in use, mainly for agriculture. Caracol was

FIGURE 6-7 LIDAR-based reconstruction of Central American Mayan city of Caracol and nearby area. *[Courtesy of Arlen and Diane Chase, Caracol Archeological Project.]*

FIGURE 6-8   Artistic reconstruction of Tenochtitlan at the site of present-day Mexico City. *[Courtesy of Tomás Filsinger.]*

but one of dozens of cities packed into this densely inhabited region. Sometime between the years 800 and 1000, the classic Mayan cultures collapsed because of some combination of droughts, environmental degradation, resource depletion, and strife among different groups.

Northeast of present-day Mexico City, a powerful culture centered on the city of Teotihuacan appeared 2,000 years ago. At the height of its power between 400 and 600, Teotihuacan boasted the third-largest stone structure in the world, the Pyramid of the Sun. Teotihuacan fell in the 700s.

By the time of European arrival, the region was under the control of an alliance of city-states headed by the Aztecs. Their capital city of Tenochtitlan was a complex of artificial islands and intervening canals that had been constructed in Lake Texcoco at the site of present-day Mexico City (Figure 6-8). Tenochtitlan was bigger than most European cities, with fresh water brought in from nearby springs, thriving wetland agriculture, and streets swept clean every night.

In summary, land clearance for agriculture was already widespread by 4,000 to 3,000 years ago in Mesoamerica because of early crop domestication. As clearance continued in most regions during subsequent millennia, and maize was domesticated, major cultures prospered. At the eve of European conquest, some fifteen million people were living in Mesoamerica. The forest clearance needed to cultivate enough land to feed fifteen million people would have emitted a substantial amount of CO_2 to the atmosphere during the interval when atmospheric CO_2 concentrations were rising (recall Chapter 3, Figure 3-7).

The Andes

The second area in the Americas with a long history of early agricultural innovation spans diverse environments in western South America—from the hyper-arid Pacific coast, across the heights of the Andes, and eastward to the warm, humid lower Andean rain forests to the east (Figure 6-9).

Local forms of squash and beans were first domesticated by 6,000 years ago in the northern Andes (present-day Ecuador). The first technologically sophisticated culture appeared between 5,200 and 4,500 years ago at an urban center called Norte Chico in coastal Peru north of modern-day Lima. Because the coast in this region receives almost no rain, it seems an unlikely place for food production sufficient to support an advanced civilization, but the ocean provided a rich supply of fish and shellfish, and the native people also learned to grow fruits and vegetables by diverting water flowing in rivers down from the Andes into terraces piled with soil.

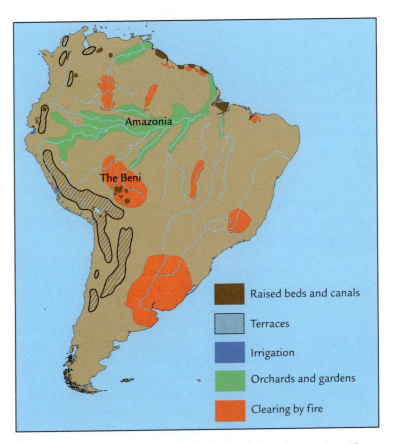

FIGURE 6-9   Centers of early agriculture in South America. *[Adapted from map by C. C. Mann, W. Doolittle, and P. Dana, in C. C. Mann* 1491 *(New York: Random House, 2006); courtesy of C. C. Mann.]*

Archeological evidence points to the early domestication of many kinds of crops across this region (see Table 6-1). On the lower eastern slopes of the central Andes, in terraced areas, peanuts appeared by 8,500 years ago, manioc (cassava or yucca) by 8,000 years ago, chili peppers by 6,000 years ago, and local varieties of beans by 5,000 years ago. In the higher central Andes, potatoes were grown by 8,000 years ago, with small early forms prior to domestication giving way to larger modern varieties (Figure 6-10). Potatoes were well suited to the Andes because they can be grown at elevations as high as 14,000 feet, despite the short frost-free seasons in this region. Chenopods (also called goosefoot and related to beets and spinaches) were also grown in high-mountain regions. By 6,000 years ago, llama and alpaca were domesticated for food and wool, but they were too small and temperamental to be suited for heavy labor.

By 2,000 to 1,000 years ago, two major cultures had developed in the Andean region: the Wari and the Tiwanaku. Near the coast and on the steep slopes of the Andes, the Wari people constructed terraces to farm potatoes, sweet potatoes, and maize. Remains of abandoned irrigated terraces are now found over an area of one million acres (Figure 6-11). The Wari were later followed by the Chimu culture, a much larger empire that grew maize and cotton near the Moche River along the Peruvian coast.

The Tiwanaku people first appeared in the central Andean highlands near Lake Titicaca around 2,800 years ago and expanded their territorial holdings around 1,500 years ago. They built canals to drain boggy areas and used raised-field agriculture to ward off late spring and early autumn frosts. As the cold mountain night air flowed over the canals, the previous day's heat stored in the water warded off freezes and extended the growing season in summer.

The Inca Empire came into being by 1450, only a few decades before European contact. From the central city of Cusco, the Inca ruled an immense and highly organized empire spanning almost 3,000 miles

FIGURE 6-11 Hillside terraces in the Andes. *[Travelpix/Alamy.]*

north to south. Without the help of livestock as draft animals or wheeled vehicles for transport, they constructed cities made of remarkable dry-wall stone masonry, using no cement, with perfectly fitted boulders of immense size. They also built dams, terraces, and irrigation canals, as well as stone-paved roads linking the entire empire, at that time the largest centrally managed road system in the world. People in the region had been working copper for centuries before the Europeans arrived, and they had begun fabricating items made of bronze. But they used metals primarily for ornamental and ritualistic purposes, while fashioning surprisingly effective armor from tightly woven cloth.

Food sources for the Inca Empire ranged from fish and irrigated crops on the Pacific coast, to potatoes in the higher terrain, to maize and manioc in the foothills and sheltered valleys, with llama and alpaca used as livestock in the higher Andes. Because much of the Inca Empire west of the Andes was not heavily forested, deforestation of this area probably contributed only modestly to the rise in CO_2 concentration occurring in later pre-industrial centuries (recall Chapter 3, Figure 3-7). However, deforestation by Incan and other people of the forested lower eastern Andes flanks likely made a more significant contribution.

Amazonia

By 11,000 years ago, early Americans in the Amazon Basin had begun cutting native rain forest trees and replacing them with a remarkable diversity of trees that produced fruits (including pineapple and peach palms) and nuts. Archeologists believe that most people lived on high bluffs near the Amazon River and its many tributaries, close enough to the water to fish, but high enough to avoid extreme seasonal flooding during the wet summer monsoon (see Figure 6-9). By 4,200 years ago, people began relying more heavily on crops, with manioc as the major staple food (Figure 6-12). Manioc appears to have been first domesticated earlier southwest of Amazonia. In addition, peach palms

had become a nutritious source of food by 2,300 years ago, and maize had arrived from the north and was planted intensively by 1,000 years ago (see Figure 6-6).

A remarkable innovation first appeared by 2,400 years ago (Figure 6-13). Farmers created **terra preta** by burning organic matter (woody debris from trees and other vegetation) in oxygen-poor conditions in order to convert wood carbon to charcoal. Adding charcoal greatly increased poor soil fertility both because beneficial microorganisms

FIGURE 6-12 Manioc, domesticated in the South American lowlands. [*Robert L. Smith.*]

FIGURE 6-13 Highly fertile terra preta soil, enriched by adding charcoal. *[Bill Woods.]*

proliferate in this kind of soil and because nutrient-rich organic matter adheres to charcoal. Some ancient plots of terra preta are "mined" today as a fertilizer source to enrich soils in other regions. Most terra preta sites were a half-meter (1–2 feet) deep and covered about 5 hectares (about 11 acres), but some were 2 meters (6 feet) deep and spanned several hundred acres.

By the time of European exploration, five to ten million people lived in the Amazon lowlands, with major urban centers near the river mouth and well inland along the river and its tributaries (recall Figure 6-9). Planting crops in these rain forest areas required substantial deforestation. Deforestation of the upper reaches of the Amazon in recent decades has uncovered many built structures indicative of high population densities and complex societies. By one estimate, at least one-eighth of the Amazon Basin had been transformed by humans before Europeans arrived.

The Beni province along the southwest margin of the Amazon Basin (in modern-day Bolivia) reveals a very different kind of anthropogenic presence (recall Figure 6-9). This region is a semi-arid savanna because of long dry winters that prevent rain forest vegetation from taking hold. But it also has six very wet summer months during the South American monsoon season, and water floods much of the landscape as it flows into the drainage system of the lower Amazon Basin.

Early people in the Beni area lived on natural topographic mounds that had been gradually built up to heights of tens of feet by the addition of earthen fill and by discarding pottery fragments and other refuse over long intervals of time. These circular islands of habitation were linked by causeways—raised earthen roads with adjacent canals several miles long. The Beni people used weirs (underwater fences) to guide and trap fish in the flooded areas, and they created raised agricultural fields that did not flood during the wet season (Figure 6-14). They also regularly set fires during the dry season to keep areas of the grass savanna clear of shrubs and trees. Farming began in the Beni region about 3,000 years ago, and the population had grown to as many as one million people by 1,000 years ago. Manioc and maize were the major crops. Like other cultures in the Amazon Basin, the people of the Beni region were well adapted to an unusually demanding environment.

FIGURE 6-14 Pre-Columbian raised-field agriculture in the Beni savanna. *[Clark Erickson.]*

North America

The areas of modern-day Canada and the United States were less densely occupied than Central America and parts of South America. Earthen mound complexes were constructed in the area that is now the state of Louisiana as early as 5,400 years ago. Several local crops were cultivated starting 4,500 to 4,000 years ago: sumpweed (also called marsh elder), chenopod (goosefoot), maygrass (also called "little barley"), and sunflowers. None of these wild plants were conducive to high food yields, with the exception of sunflowers. Only two early crops, sunflowers (with their large seeds) and eastern varieties of squash (such as acorn squash and several summer squashes) have remained in use through the present day. Between 2,500 and 2,100 years ago, the Adena culture developed along the Ohio River and influenced other peoples as far northeast as modern-day New England (Figures 6-15 and 6-16). The Adena built small burial mounds and farmed the crops noted above, as well as tobacco.

Around 2,200 years ago, the Hopewell culture developed in the Ohio River valley and in woodlands to the west. These people grew the same crops as the Adena. Maize had begun to spread north from Mexico, but could not yet fully flourish in colder climates farther north.

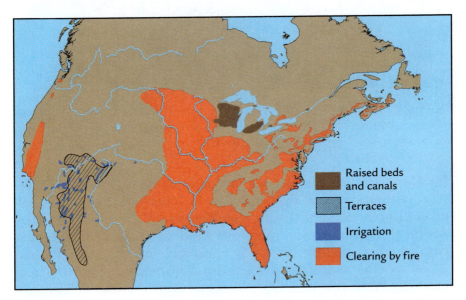

FIGURE 6-15 Centers of early agriculture in North America. *[Adapted from map by C. C. Mann, W. Doolittle, and P. Dana, in C. C. Mann, 1491 (New York: Random House, 2006); courtesy of C. C. Mann.]*

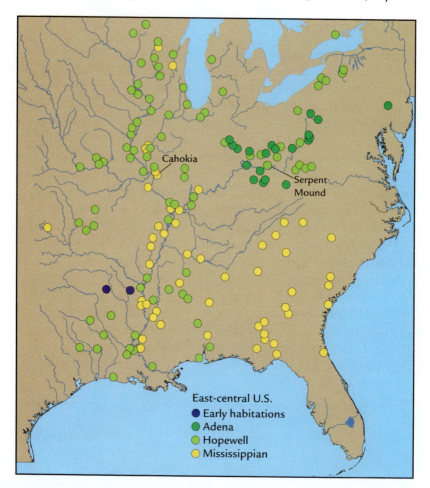

FIGURE 6-16 Prominent earthen mounds in North America built by the Adena, Hopewell, and Mississippian cultures. *[Adapted from map by C. C. Mann, W. Doolittle, and P. Dana, in C. C. Mann, 1491 (New York: Random House, 2006).]*

The Hopewell used soil to build monumental mounds for burials and ritual celebrations, and they traded with other river-dwelling people across east-central North America (see Figure 6-16). In arid southwestern North America, maize and squash of Mexican origin were planted in irrigated river valleys by 2,000 years ago, and beans by 1,000 years ago.

Between the years 200 and 900, milpa crops began to be farmed more intensively in the east, including varieties of maize that had by then been adapted to cooler climates. With the arrival of beans by the year 1000, the combined nutritional package from maize, beans, and squash spurred the development of the Mississippian culture in southeastern North America

FIGURE 6-17 Artist's reconstruction of Cahokia mounds near St. Louis, Missouri. [*Courtesy of Cahokia Mounds State Historic Site, painting by William R. Iseminger.*]

(see Figure 6-16). These people also encouraged the growth of a wide variety of nut-bearing trees (pecans, hickory, chestnuts, walnuts, butternuts, hazelnuts, and oaks) through forest management (cutting some trees to favor others) and burning.

The Mississippian people built urban centers with huge mounds topped by flat platforms where rulers lived and rituals were performed. The largest of these cities was Cahokia, just across the Mississippi River from modern-day St. Louis, with a central earthen mound 30 meters (approximately 100 feet) high, 300 meters (approximately 1,000 feet) long, and 215 meters (approximately 700 feet) wide (Figure 6-17). Cahokia was devastated by flooding and earthquakes around 1250, but villages in other mound areas remained in existence. In many regions, earthen structures called effigy mounds were fashioned to resemble animals, including the 1,330-foot-long Serpent Mound still found in modern-day Ohio (Figure 6-18).

In 1539, the Spanish explorer Hernando de Soto arrived in Florida and moved north through what is now Georgia, Tennessee, South Carolina, Alabama, the lower Mississippi River valley, and the rivers of southeast Arkansas, Oklahoma, and Texas. He found the land "thickly inhabited," with the valleys of the Tennessee and Mississippi

FIGURE 6-18 Serpent Mound in present-day Ohio. [*Richard A. Cooke/CORBIS.*]

river systems dense with villages (almost within eyesight of each other), and crops of maize and beans grown on bluffs and natural levees above the lower floodplains. Several million people probably lived in North America at this time.

Farther east, the forests were described by European settlers as remarkably clear of understory vegetation because of repeated burning by early Americans. The terrain that was kept clear by fire hosted deer, turkey, and elk, as well as herds of buffalo (bison) roaming from present-day upstate New York in the north to present-day Georgia in the south. Some historical reports describe most of the Shenandoah Valley of Virginia, a region of natural forest, as having been open grassland maintained by frequent burning. "Buffalo" is a common historic place name in the Shenandoah Valley as well as elsewhere in the east. Apparently, grassy corridors opened by man-made fires had allowed herds of buffalo to spread east after 1,000 years ago. The last of these creatures in the eastern United States were killed around 1800.

Summary

When Europeans arrived in the Americas, they encountered a population estimated at forty to sixty million people, or about 10% of the global population at the time. Most early Americans, prior to the arrival of Europeans, were farmers who cleared forests to establish cultivated plots. Some of these people maintained their villages of residence for long intervals of time, while others moved frequently. Lacking livestock (except the llama and alpaca of the Inca), early Americans used fire to manage landscapes, thereby attracting game and promoting growth of berries, nuts, and other foods. The extent of open-field burning remains uncertain. In any case, carbon emissions from forest clearance by these growing populations of early Americans would have contributed substantially to the observed rise in CO_2 during the last several millennia.

Methane-generating activities were much smaller in the Americas than in Eurasia. CH_4 emissions from llama and alpaca were insignificant, and wet-rice irrigation did not exist. However, emissions from repeated and widespread burning of grasslands and forest understory vegetation would have contributed to the observed CH_4 increase during the last 5,000 years (recall Chapter 2, Figure 2-8).

Additional Resources

Chase, A. F., D. Z. Chase, J. F. Weishampel, J. B. Drake, R. L. Shrestha, K. C. Slatton, J. J. Awe, and W. E. Carter. "Airborne LiDAR, Archaeology, and the Ancient Maya Landscape of Caracol, Belize." *Journal of Archaeological Science* 38 (2011): 387–398, doi:10.1016/j.jas.2010.09.018.

Denevan, W. M., ed. *The Native Population of the Americas in 1492.* Madison: University of Wisconsin Press, 1978.

Denevan, W. M. "The Pristine Myth: the Landscape of the Americas in 1492." *Annals of the Association of American Geographers* 82 (1992): 369–385.

Krech, S., III. *The Ecological Indian.* New York: W. W. Norton, 1999.

Lentz, D. L., ed. *Imperfect Balance: Landscape Transformations in the Pre-Columbian Americas.* New York: Columbia University Press, 2000.

Mann, C. C. *1491.* New York: Random House, 2006.

Roosevelt, A. C. *Moundbuilders of the Amazon: Geophysical Archaeology on Marajo Island, Brazil.* San Diego: Academic Press, 1991.

Sauer, C. O. "A Geographic Sketch of Early Man." *Geographical Review* 34 (1944): 529–573.

CHAPTER 7

AFRICA, AUSTRALIA, AND OCEANIA

P re-industrial humans had widely varying effects on land clearance in three additional areas: the continents of Africa and Australia, and the group of islands in the southwest and equatorial Pacific collectively known as Oceania. Africa, the continent where the human species originated, has a rich history of land clearance, agriculture, and tending of livestock that varies greatly from its northern border on the Mediterranean Sea to its narrow southern tip, primarily because of the wide range of climate across those 7,000 kilometers. In Australia, pre-industrial people never made the transition to traditional agriculture, but they transformed the landscape by setting fires, as did the much more recent arrivals in New Zealand. The people that slowly moved out of southeastern Asia and across the Pacific islands of Oceania in recent millennia adapted and domesticated many types of plants for food and had widely varying effects on the landscapes of those islands.

Africa

After millions of years of evolution, hominids with fully modern characteristics appeared in Africa between 200,000 and 100,000 years ago. DNA evidence reveals several subsequent waves of migration out of Africa into other continents, as described in previous chapters. The last major migration, involving people with modern cultural capabilities, occurred near 50,000 years ago, with people spreading into Asia, Europe, Australia, and later the Americas.

Africa is a very large continent (about 20% larger than North America and half the size of Eurasia) and extends from 35°N latitude to 35°S. It spans several distinct climate and vegetation zones: a narrow Mediterranean northern fringe with wet winters and dry summers; the immense, year-round arid Sahara Desert just to the south; the mixed tree and grass savanna of the Sahel region where rain falls mainly in summer; the year-round wet tropical rain forests near the equator; and the semi-arid tree and grass savanna that spans much of Africa south of the equator (Figure 7-1). Most of Africa lacks a written history, but the story of its agriculture can be traced through archeological data and the spread of distinctive languages (Table 7-1).

Mediterranean and Saharan North Africa

Northernmost Africa—consisting of modern-day Algeria, Morocco, Tunisia, northern Libya, and Egypt—shares climatic attributes with southern Europe and the eastern Mediterranean coast. It also shares cultural affinities with these regions because of the strong historical trading links that developed across the Mediterranean Sea. Carthage

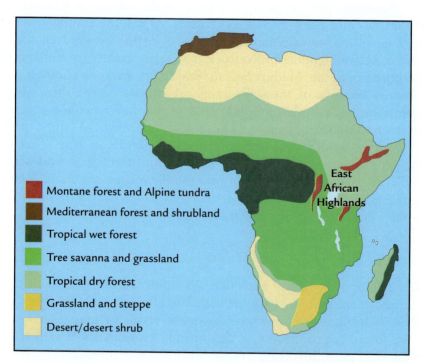

FIGURE 7-1 Major vegetation groups in Africa. [*Adapted from R. L. Smith and T. M. Smith,* Elements of Ecology *(Menlo Park: Benjamin/Cummings, 1998).*]

TABLE 7-1 **Crop and Livestock Domestication in Africa**

Crops	Time Domesticated	Region
Pearl millet	By 4,500 years ago	Sahel (western Sub-Sahara)
Sorghum	By 4,000 years ago	Sahel (eastern Sub-Sahara)
Cow peas	By 3,700 years ago	Sahel (western Sub-Sahara)
Beans	By 3,700 years ago	Sahel (western Sub-Sahara)
African rice	By 2,000 years ago	Sahel (western Sub-Sahara)
Teff	By 4,000 years ago	East African Highlands
Finger millet	By 4,000 years ago	East African Highlands
White yam	By 5,000 years ago	Western north tropical
Kola nuts	By 5,000 years ago?	Western tropical
Oil palm	By 5,000 years ago?	Western tropical

Livestock	Time Domesticated	Region
Donkey	7,000–5,000 years ago	Egypt

(modern Tunisia) was active in international trade by 2,800 years ago, and the Carthaginian Empire reached its apex 2,300 years ago. Initially a strong rival to Rome, Carthage was eventually conquered, and its land was used by the Roman Empire as a source of lumber and food. Crops and livestock domesticated in the Fertile Crescent prior to 9,000 years ago spread westward along the Mediterranean coast to this region (Figure 7-2). For a time, North Africa was called the "granary of Rome." Because of the limited extent of North African forests, early deforestation in this region was likely only a minor contributor to global CO_2 emissions.

In northeastern Africa (Egypt), crops like wheat, barley, and chickpeas arrived from the Fertile Crescent by 9,000 to 8,000 years ago, along with domesticated cattle, sheep, and goats (see Figure 7-2). Later, the donkey was domesticated in this region. Life in Egypt centered on the Nile River, which originates in highland sources far to the south and flows north through hyperarid desert (Figure 7-3). Because agriculture in this area was totally reliant on Nile water, dense populations grew near the river, resulting in an agricultural economy by 7,500 to

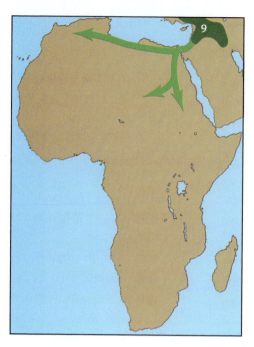

FIGURE 7-2 Dispersal of Fertile Crescent agriculture into Mediterranean North Africa and the Nile River valley. [Adapted from P. Bellwood, First Farmers: The Origins of Agricultural Societies (Oxford: Blackwell, 2004); and M. D. Purugganan and D. Q. Fuller, "The Nature of Selection During Plant Domestication," Nature 457 (2009): 843–848, doi:10.1038/nature07895.]

7,000 years ago, and an organized political state by 6,000 years ago. The Nile was also the pathway along which Fertile Crescent agriculture and livestock domestication penetrated southward into the highlands of Ethiopia.

In recent millennia, the harsh barrier of the Sahara Desert has limited north–south contact among African populations, but between 10,000 and 5,000 years ago, much of the Sahara was not the hyperarid area of today. At that time, stronger summer monsoons sent rains much farther north (see Chapter 2, Figure 2-5), creating a grass-covered landscape dotted with lakes and seasonal waterways. The lower and wetter regions hosted hippos, crocodiles, rhinos, and turtles, while the intervening grasslands held gazelles, hares, and other game.

Early on, pastoral people grazed livestock on these grasslands, including cattle prior to 7,500 years ago and, later, sheep and goats as well. These people lived nomadic lives, constantly moving in search of water and grass and slowly dispersing westward from the Nile region across North Africa. The population densities of these nomadic groups were never high because of the extreme aridity of the environment, and the contribution of their livestock to the atmospheric methane trend was likely negligible.

FIGURE 7-3 Fertile lands around the Nile River running through a desert landscape. *[Andia/Alamy.]*

The Sahel and East African Highlands

As the summer monsoon weakened after 5,000 years ago and no longer reached as far northward into the Sahara, herders followed the retreating rains southward to the grass and tree savanna of the Sahel (Figure 7-4) as well as the highlands of East Africa. As they moved south, they introduced livestock to these areas for the first time (Figure 7-5). The growing livestock herds would likely have contributed to the rise in atmospheric methane concentrations after 5,000 years ago (recall Chapter 2, Figure 2-8).

People in the southern Sahel also began to domesticate several crops. In the west-central Sahel, pearl millet (a cereal grain) was domesticated by 4,500 to 4,000 years ago, cow peas by 3,700 years ago, and African rice by 3,000 to 2,000 years ago (Figure 7-6). In the eastern Sahel, sorghum (both a cereal crop and a sweetener like molasses) was domesticated by at least 4,000 years ago (Figure 7-7). With archeological records from the Sahel very sparse, it is likely that these crops were domesticated earlier, as indicated by the fact that pearl millet and sorghum of African origin are found in India by 4,500 to 3,500 years ago. Domestication of these crops in Africa may also have occurred farther north than the current limit of the

A

B

FIGURE 7-4 Examples of (A) an African savanna (mixed trees and grasslands) and (B) an African rain forest. *[A: Richard Cooke/Alamy; B: NHPA/SuperStock.]*

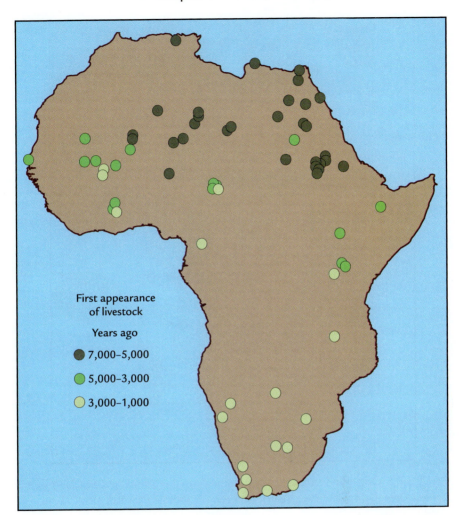

FIGURE 7-5 The spread of domesticated livestock across Africa. *[Adapted from D. Q. Fuller et al., "The Contribution of Rice Agriculture and Livestock to Prehistoric Methane Levels: An Archeological Assessment," The Holocene 25 (2011): 743–759, doi:10177/ 0959683611398052.]*

Sahel, because the summer monsoon rains were stronger then than today.

As the monsoon slowly retreated south, and as agricultural clearance increased, the intermittent savanna tree cover across the Sahel was gradually reduced. The northern part of the East African Highlands followed a similar path to the Sahel. Cattle and goats from southwest Asia appeared in Kenya by 6,000 to 5,000 years ago and a form of millet was domesticated in Ethiopia by 4,000 years ago.

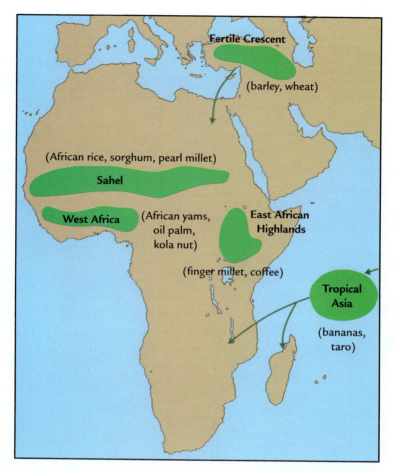

FIGURE 7-6 Origin and spread of food domesticated in Africa. *[Adapted from J. Diamond,* Guns, Germs, and Steel: The Fates of Human Societies *(New York: W. W. Norton, 1999).]*

Tropical Africa

In Africa, rain forests (see Figure 7-4b) occupy a large region near the equator that includes the watersheds of the Niger and Congo Rivers. Very early farmers in this region used polished stone axes, as they did in Europe. By 4,000 years ago, people in the African tropics were producing tools made of copper, and iron technology developed by 2,600 years ago. Some archeologists have suggested that the high-temperature furnaces in this region were capable of producing steel-like iron that was far superior to that used in Mediterranean Europe.

These rain forests, which could otherwise have been a productive agricultural area, have long escaped much of the deforestation typical of other tropical regions because a stew of tropical diseases kept populations in check, including malaria (transmitted by mosquitoes, and resulting in debilitating fevers and often death), and trypanosomiasis (transmitted by tsetse flies, and resulting in sleeping sickness in both cattle and humans).

Dates of food domestication in the African tropics are uncertain, because highly corrosive rain forest soils preserved few archeological records. Early farming (by around 5,000 years ago) focused mostly on root crops and nuts from trees. Foods domesticated just north of the equator included yams, kola nuts, and oil palm (Figures 7-8 and 7-9). Later, between 2,000 and 1,000 years ago, ships from southeastern Asia introduced taro, bananas, and other foods to tropical Africa.

The Bantu people of western tropical Africa grew yams and harvested kola nuts and oil palms in the area of modern-day Cameroon and Nigeria, and in time their agricultural success led to population growth. After 5,000 years ago, the expanding populations began to move southward and displace many of the pygmy peoples that had been living in the tropical African rain forests (Figure 7-10). As the Bantu population increased, they cut clearings in the rain forest, though probably at first without any large contribution to CO_2 emissions on a global scale.

Between 4,000 and 3,000 years ago, the Bantu also spread eastward into the East African Highlands, where people from the Sudan area to the north had previously introduced livestock (mostly goats and cattle), as well as fowl (guinea hens), and finger millet (a cereal). As the Bantu added these new practices, along with sorghum farming, to their basic root-crop farming, their populations grew even larger (see Figure 7-10).

FIGURE 7-8 Yams, domesticated in tropical North Africa. *[Robert L. Smith.]*

Ecologists and archeologists have identified an interval between 2,800 and 2,200 years when the central African rain forest became markedly less dense. This thinning has been widely attributed to an interval of dry weather, but an ocean sediment core located just offshore from the Congo River contains information that suggests a different interpretation. Two geochemical indices in that core show increases that would normally indicate weathering of soils under a wetter, not a drier, climatic regime. This evidence makes the dry-climate explanation for forest thinning suspect.

Another possible explanation for the thinning rain forest of this area is that the Bantu people had by that time cleared enough forest and farmed enough land to expose soils to greater physical weathering. With more freshly eroded debris lying around on the landscape, all of the products of weathering would have increased, even without any change in rainfall. If the explanation of increased clearance turns out to be correct, the interval of forest thinning that began 2,800 years ago in central Africa may have been a time of growing anthropogenic CO_2 emissions to the atmosphere.

Subtropical Southern Africa

South of the central African rain forest, the natural landscape of southern Africa is a diverse mixture of semi-arid grasslands, semi-humid woodlands, and other vegetation intermediate between these extremes (see Figure 7-1). For many thousands of years, the

FIGURE 7-9 Taro, domesticated in southeast Asia and prominent among Bantu crops in western tropical Africa. *[Robert L. Smith.]*

FIGURE 7-10 Dispersal of Bantu people and agriculture from origins in tropical Africa. Numbers show thousands of years ago. *[Adapted from P. Bellwood,* First Farmers: The Origins of Agricultural Societies *(Oxford: Blackwell, 2004); and M. D. Purugganan and D. Q. Fuller, "The Nature of Selection During Plant Domestication,"* Nature *457 (2009): 843–848, doi:10.1038/nature07895.]*

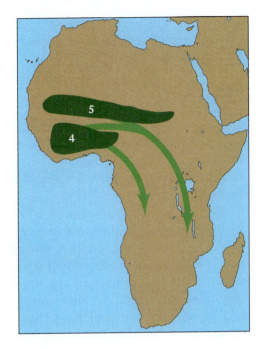

Khoisan (Khoi and San) people in this region lived mainly by hunting, gathering, and fishing, with domesticated livestock arriving around 3,000 to 2,000 years ago (see Figure 7-5).

Then, between 3,000 and 1,000 years ago, the Bantu people expanded southward and introduced their mixture of crops (yams, taro, millet, and sorghum) and livestock (see Figure 7-10). The Bantu typically moved first into wetter regions where yams and taro would grow, and then slowly expanded out into drier regions, displacing the Khoisan. This "Bantu expansion," which carried populations 3,500 kilometers to the south in less than 1,000 years (with much of the migration occurring between 2,100 and 1,600 years ago), was one of the most remarkable agricultural and human migrations in history. It occurred because Bantu agricultural practices were more successful in feeding their people than the Khoisan's hunter-gatherer practices.

Within a few centuries, the Bantu came to dominate most of southern Africa, with only small pockets of Khoisan remaining. The southward movement of Bantu agriculture was checked only by the climatic restrictions of the year-round hyperaridity of the Kalahari Desert in the southwest (see Figure 7-1) and the different rainfall pattern in far southern Africa, where precipitation occurred in winter. The Bantu expansion contributed additional CO_2 emissions during the last 5,000 years because of deforestation, as well as CH_4 emissions from the spread of livestock (see Figure 7-5).

Australia

Australia is the smallest, flattest, and driest continent. Its forests are restricted mainly to the tropical (monsoonal) northern region, the area of modern-day Queensland, and a narrow strip along its eastern coast, modern-day New South Wales and Victoria, with grasslands and deserts everywhere else (Figure 7-11).

Around 50,000 years ago, humans first arrived in Australia, probably after gradually spreading eastward from Africa by way of coastal Arabia and India, and then southward along the islands of southeast Asia. Although these Aboriginal people were hunter-gatherers, many scientists think they had a large impact on the Australian landscape. They used "firesticks" to burn forests, open up grasslands, and drive wild game, most of which in this isolated continent consisted of marsupials (mammals that carry and nurse their young in stomach pouches). In *The Fire Eaters*, Tim Flannery suggests that by using fire as a form of "game management," Aboriginal people caused a

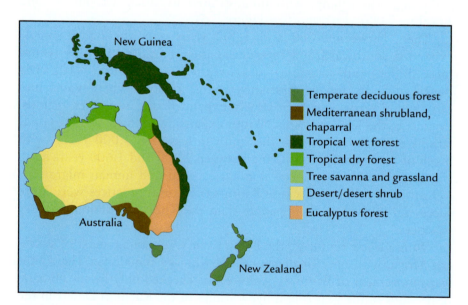

FIGURE 7-11 Major vegetation groups in Australia. *[Adapted from R. L. Smith and T. M. Smith,* Elements of Ecology *(Menlo Park: Benjamin/Cummings, 1998).]*

major rearrangement of the natural landscape, including the larger animals.

Prior to the arrival of the Aboriginal people, a diverse population of marsupials had lived on the continent but, within a relatively short time, many of these had gone extinct. As happened in the Americas (recall Chapter 6), most of the species that disappeared were larger in size than humans, including a marsupial lion, three kinds of wombat, nine kinds of kangaroo, a giant lizard, a giant tortoise, an ostrich-sized flightless bird, and others. Climate was not changing dramatically at the time and is unlikely to have been the cause of these widespread extinctions. The evidence points to humans as the likely culprit. Having evolved without humans present, the marsupials probably had no natural fear and were easy prey for hunters, who also set habitat-destroying fires to drive game into specific areas and make them easier to hunt.

Early Australians never developed full-fledged agriculture, in large part because their native plants were not easily suited for domestication. People did gather, thresh, and grind wild millet, but they didn't take the next step of planting it as a crop. However, it appears that some Aboriginal people "farmed" eels in canals constructed for that purpose, and their use of fire also stimulated growth of edible root crops.

In addition to killing off much of the marsupial population, early burning of the land caused a reduction of Australia's forests, but it seems unlikely that the increased burning around 45,000 years ago caused any significant increase in atmospheric greenhouse-gas levels, because Australia is small and arid enough that it never had really large forests. As a result, no distinctive signal is likely to have been left by Australian human activities in the ice-core CO_2 or CH_4 records from 45,000 years ago. In more recent times, prior to the arrival of Europeans in the late 1700s, the Aboriginal population remained small (a few hundred thousand people). True agriculture still had not developed, so greenhouse-gas emissions by humans were not large.

Oceania

Most of the western Pacific islands close to Eurasia were first inhabited between 50,000 and 30,000 years ago, as part of the migrations of the hunting-fishing-gathering people described previously. One island in this region—New Guinea—has an agricultural history distinctly different from the others. In contrast to nearby arid Australia, tropical New Guinea is wet, covered by forests, and drained by rivers (see Figure 7-11). By 9,000 years ago, people were building drainage ditches to shed heavy tropical rainfall from wetter areas and constructing terraces to retain

TABLE 7-2 **Plant Domestication in New Guinea**

Plant	Time Domesticated	Region
Taro	9,000(?)–7,000 years ago	Highlands
Yam	9,000(?)–7,000 years ago	Lowlands
Banana	9,000(?)–7,000 years ago	Highlands
Sugar cane	9,000(?)–7,000 years ago	Highlands

water in drier areas. By 6,800 years ago, New Guinea had become an independent center of agriculture based on root crops like yam and taro, as well as banana and sugar cane (Table 7-2).

The beginning of agriculture in New Guinea was unusual; it developed in the central highlands at elevations above 1,300 meters (4,000 feet), rather than along the coasts. The success in growing taro and sweet potatoes resulted in very dense Neolithic populations, and a total population of perhaps one million people. As a result, the highlands were not overrun by later arrivals of people from other areas, and even now the highlanders retain the darker pigmentation of their ancestors from 50,000 to 30,000 years ago. Subsequent arrivals from the mainland near 3,600 years ago introduced pigs and chickens to New Guinea.

Elsewhere in Oceania, island people closer to Asia adopted many aspects of mainland agriculture and fishing. People in what is now Taiwan (off the coast of China), and the Philippines farther to the south, took up rice irrigation by 5,000 to 4,000 years ago, while people in Borneo and Indonesia did so between 4,000 and 3,000 years ago (see Chapter 5, Figure 5-14).

People from southeast Asia and tropical islands offshore then spread across Fiji and Samoa around 3,000 years ago, bringing taro, yams, bananas, coconuts, and breadfruit, as well as pigs, chickens, and dogs from China and the southeast Asian islands. A later wave of migration of highly skilled navigators carried people far out into the Pacific— northeast to Hawaii, east to Easter Island, and south to New Zealand (Figure 7-12).

New Zealand

New Zealand provides an unusually distinct and well-understood example of how humans can affect natural landscapes. Within a few hundred years of the arrival of the Maori people, around the year 1280, the native bird population suffered a wave of extinctions that comprised

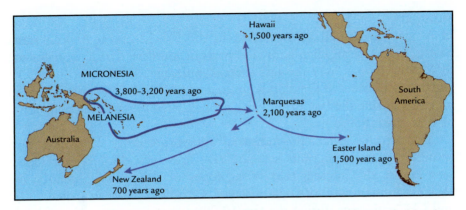

FIGURE 7-12 First human occupation of the widely separated oceanic islands of the Pacific. (Recent evidence suggests that some dates may be younger than shown here.) *[Adapted from J. Diamond,* Guns, Germs, and Steel: The Fates of Human Societies *(New York: W. W. Norton, 1999).]*

ten species of moa, including giant flightless birds that stood as much as seven feet tall, as well as the giant short-winged eagle, and countless species of smaller birds.

Prior to human arrival, charcoal records from lake sediments show that fire was rare in the relatively moist New Zealand climate, but the Maori began burning forests to encourage the growth of understory bracken (ferns) with roots that provided a nutritious source of food. As each region was burned for the first time, a large pulse of charcoal was deposited in lake sediments for a few decades. These charcoal-rich sediments appeared in easily accessible coastal regions soon after 1280, but not until as late as 1450 in remote interior areas. After initial clearance, charcoal deposition fell to much lower levels as the Maori continued to set much smaller fires to keep the forests from returning.

By most estimates, almost half of New Zealand's forests were destroyed in this way. A population estimated at just 5,000 people managed to deforest much of South Island. For the total New Zealand land area of 265,000 square kilometers, and an estimated total population of not much more than 100,000 people, the average clearance per person was a remarkably high 100 hectares (roughly 220 acres). New Zealand provides unambiguous evidence of a Neolithic culture, lacking advanced technology and with relatively low population densities, using fire to cause remarkably extensive clearance. Studies elsewhere in the western and equatorial Pacific show that major deforestation occurred on drier islands when humans first arrived.

Summary

In Africa, substantial deforestation in the Sahel, central African rain forests, East African Highlands, and parts of southern Africa contributed to CO_2 emissions over the last several millennia. Tending of livestock and yearly burning of grasslands emitted CH_4. Rapid population growth in Africa during the last 1,000 years probably led to increased emissions of both greenhouse gases.

In Australia, where pre-industrial agriculture never developed, the earliest human immigrants set extensive fires near 45,000 years ago, long before the interval that is the focus of this book. Similarly, in Oceania human activities again played a negligible role in global CO_2 and CH_4 emissions during the last 10,000 years, because the total land area and the human populations were small. But both Australia and New Zealand offer instructive examples of ways in which early humans interacted with landscapes, including extensive forest clearance at low population densities.

Additional Resources

Bayon, G., B. Dennielou, J. Etoubleau, E. Ponzevera, S. Toucanne, and S. Bermell. "Intensifying Weathering and Land Use in Iron Age Central Africa." *Science* 335 (2012): 1219–1222. doi:10.1126/science.1215400.

Diamond, J. *Guns, Germs, and Steel: The Fates of Human Societies.* New York: W. W. Norton, 1999.

Ehret, C. *An African Classical Age: Eastern and Southern Africa in World History, 1000 B.C. to A.D. 400.* Charlottesville: University Press of Virginia and Oxford: James Currey, 1998.

Flannery, T. *The Future Eaters: An Ecological History of the Australasian Lands and People.* New York: Braziler, 1998.

McGlone, M. S., A. J. Anderson, and R. N. Holdaway. "An Ecological Approach to the Polynesian Settlement of New Zealand," in *The Origins of the First New Zealanders,* edited by D. G. Sutton. Auckland: Auckland University Press, 1994.

McWethy, D. B., C. Whitlock, J. M. Wilmshurst, M. S. McGlone, and X. Li. "Rapid Deforestation of South Island, New Zealand, by Early Polynesian Fires." *The Holocene* 19 (2009): 883–897. doi:10.1177/0959683609336563.

Miller, G. H., J. W. Magee, B. J. Johnson, M. Fogel, N. A. Spooner, M. T. McCullough, and L. K. Ayliffe. "Pleistocene Extinction of *Genyornis newtorii*: Human Impact on Australian Megafauna." *Science* 283 (1999): 205–208.

Rolette, B., and J. Diamond. "Environmental Predictors of Pre-European Deforestation on Pacific Islands." *Nature* 431 (2004): 443–446. doi:10.1038/nature02801.

Part 2 Summary

During the warmth of the current interglaciation, our farming ancestors first domesticated crops and livestock from plants and animals that were naturally available in the regions where they settled (Asia, the Americas, and Africa). Domestication of some major food sources began as early as 10,000 years ago, but in other cases did not occur until thousands of years later (Figure 2s-1). Many dates of first domestication remain uncertain.

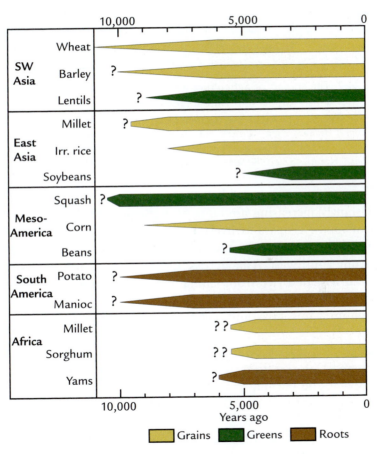

FIGURE 2s-1 Time of domestication of major food-producing crops in Asia, Africa, and the Americas. *[Adapted from P. Bellwood,* First Farmers: The Origins of Agricultural Societies *(Oxford: Blackwell, 2004); and M. D. Purugganan and D. Q. Fuller, "The Nature of Selection During Plant Domestication,"* Nature *457 (2009): 843–848, doi:10.1038/nature07895.]*

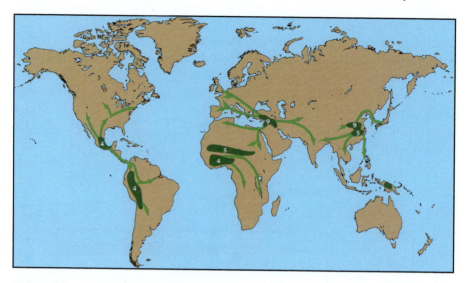

FIGURE 2s-2 The spread of agriculture across the continents during pre-industrial time. Numbers in white indicate the time of initial spread, in thousands of years ago. Light-green arrows show the major pathways.
[Adapted from P. Bellwood, First Farmers: The Origins of Agricultural Societies (Oxford: Blackwell, 2004); and M. D. Purugganan and D. Q. Fuller, "The Nature of Selection During Plant Domestication," Nature 457 (2009): 843–848, doi:10.1038/nature07895.]

Constantly increasing amounts of archeological data now map the spread of major crops and livestock across Earth's surface during the last several thousand years, including wheat, barley, goats, sheep, and cattle from southwest Asia across Europe and Southern Asia; pigs, millet, and irrigated rice from China across eastern and southern Asia; corn (maize), squash, potatoes, beans, and tomatoes across the Americas; and sorghum, yams, taro, and millet across tropical and southern Africa (Figure 2s-2).

This gradual expansion of farming emitted greenhouse gases into the atmosphere. Deforestation produced CO_2, while tending livestock, irrigating rice, and burning grasses and shrubs produced CH_4. The timing of the spread of agriculture across the continents fits into the interval of rising CO_2 and CH_4 trends of the last few thousand years that were described in Part 1. This correlation in time suggests that human activities played a role in the wrong-way greenhouse-gas trends during pre-industrial times. Part 3 will take the next step of trying to quantify the size of that role.

Debating a New Hypothesis

I n 2003, the wrong-way greenhouse-gas trends (reviewed in Part 1) and the available evidence about the spread of agriculture (reviewed in Part 2) formed the basis for a new idea called the **early anthropogenic hypothesis.** I proposed that the spread of agriculture during the last several thousand years had caused increasing greenhouse-gas emissions that could account for the wrong-way gas trends. I also claimed that the gas released from agriculture during the last several thousand years kept Earth's climate warmer than it would have been had Nature stayed in control. This proposed link between agriculture and the wrong-way gas trends was presented to a broader audience in an article published in *Scientific American* in 2005.

The real world of science is not the quaint ivory-tower pursuit of pure knowledge some non-scientists imagine it to be; it is instead a competitive and at times confrontational endeavor. As new ideas emerge, scientists are not the least bit shy about challenging them, especially if these ideas are provocative enough to have implications that range across several fields of knowledge.

Many new hypotheses are quickly rejected and vanish from sight because they fail to pass subsequent testing by the scientific community, but the early anthropogenic hypothesis has not been one of those. A decade after its initial appearance, it has received attention from, and been tested by, scientists in an expanding range of disciplines, including archeology and archeobotany, land-use modeling, astronomy, ice-core and marine geochemistry, and carbon and climate modeling. In addition, historical records have been brought into the debate. In light of all this attention, the early anthropogenic hypothesis has at least passed the first hurdle of basic survival.

Each chapter in Part 3 summarizes evidence, both for and against the hypothesis, that has emerged during the decade that it has been debated. Has the early anthropogenic hypothesis successfully met the challenges posed by its critics and reached the point that it merits larger scientific acceptance, or has it failed those challenges and should thereby be rejected? The natural explanations proposed for the greenhouse-gas trends of the last several thousand years deserve equal consideration: how well have they stood up to critical examination?

Chapter 8 explores an initial reaction common to many scientists: the belief that too few people lived thousands of years ago to have had more than a small effect on atmospheric greenhouse-gas trends, even though agriculture was continuously spreading across Earth's surface (recall Part 2). Land-use modelers investigating this issue have generally assumed that farmers have always used the same average amount of land per person, but a range of historical and archeological evidence surveyed in Chapter 8 disagrees with this assumption.

Chapter 9 examines a challenge to the early anthropogenic hypothesis that involves the alignment of greenhouse-gas trends in the current and previous interglaciations. The alignments presented in Part 1 were based on independently calculated changes in Earth's orbit, but other methods have been suggested. What do alternative alignment methods reveal about gas trends during prior interglaciations? Did gas concentrations rise (as in the current interglaciation) or did they fall (marking this interglaciation as anomalous)? Chapter 9 also adds new information on CO_2 and CH_4 trends during four earlier interglaciations for which ice sequences had not yet been recovered when the early anthropogenic hypothesis was published in 2003.

Chapter 10 expands the initial analysis of methane trends summarized in Chapter 2. It explores whether or not natural sources of this gas in the Southern Hemisphere might account for the upward CH_4 trend during the last 5,000 years, and it summarizes the first quantitative estimates of methane emissions from human activities (specifically, irrigated rice farming) during those millennia.

Chapter 11 expands the summary of carbon dioxide trends in Chapter 3 and attempts to evaluate whether natural or human-generated emissions best account for the upward CO_2 trend that began 7,000 years ago. It reviews quantitative estimates of carbon emissions based on differing assumptions about the amount of land use by farmers in the past. It also assesses other factors that may have affected anthropogenic carbon emissions: burning coal and peat, feedback from the ocean, the unusually large early use of fire to clear land, and the effect of early population estimates higher than those normally used. Chapter 11 also explores geochemical constraints on net CO_2 emissions.

Additional Resources
Ruddiman, W. F. "The Atmospheric Greenhouse Era Began Thousands of Years Ago." *Climatic Change* 61 (2003): 261–293.
Ruddiman, W. F. "How Did Humans First Alter Global Climate?" *Scientific American* (March 2005): 46–53.

CHAPTER 8

EARLY FARMING AND PER CAPITA LAND USE

From the start, the early anthropogenic hypothesis faced a widely voiced challenge: how could the relatively small number of farming people living thousands of years ago have transformed enough of Earth's surface to emit large amounts of greenhouse gases? This question highlights the fact that global population underwent its greatest increase only within the last few hundred years, from about 450 million people in the year 1500 to 900 million people at the dawn of the Industrial Era in 1850, and then to more than 7 billion people today. In contrast, the pre-industrial concentrations of both greenhouse gases (and particularly CO_2) began to increase thousands of years ago, when the estimated world population was much smaller (Figure 8-1).

Part 2 of this book showed that early farming activities first expanded thousands of years ago, about the same time that greenhouse-gas concentrations reversed their downward trend and began to rise. Early forest clearance on a large scale began in Europe and China near the start of the CO_2 increase 7,000 years ago (recall Chapters 4 and 5). Early rice irrigation in China and a major spread of domesticated livestock in Asia and Africa occurred near the time of the reversal in the CH_4 trend 5,000 years ago (recall Chapters 5 and 7). These concurrences suggest a link between human activities and greenhouse-gas concentrations, but major rises in greenhouse gas concentrations

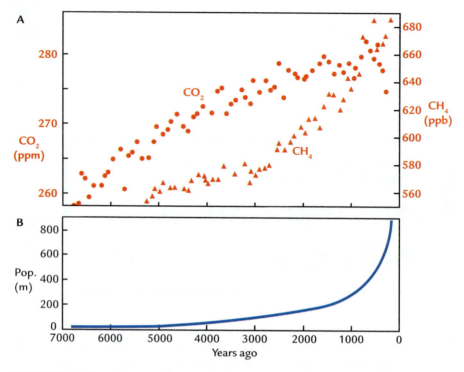

FIGURE 8-1 Trends in greenhouse-gas concentrations (A) and estimated global population (B) during the last 7,000 years. *[Gas concentrations (A) from EPICA Community Members, "Eight Glacial Cycles from an Antarctic Ice Core," Nature 429 (2004): 623– 628. Populations (B) based on C. McEvedy and R. Jones, Atlas of World Population History (New York: Penguin Books, 1978).]*

preceded the much later population increase, raising the question of whether farming was really more than just a minor factor in the early gas rises.

Model-based attempts to reconstruct land use prior to the last few centuries have been underway for more than a decade, and the scientists doing this work have generally relied on a key simplifying assumption—that the amount of forest cleared was linked to past populations on a one-to-one basis (or close to it). This assumption is often justified by the question: "If each farming family needed only a certain amount of land to feed themselves, why would they have farmed more than that amount?" In this modeling approach, modern clearance and population levels are first linked in a quantitative relationship, and that relationship is then used to predict past clearance based on past populations.

But is this approach really as reasonable as it initially sounds? Consider another equally reasonable-sounding question: "Have farmers really been so dull-witted that they learned nothing new about their basic livelihood during the last 7,000 years, not a single new skill that would enable them to get more food from the land they farmed?" From this perspective, the assumption of constant land use in the past sounds unlikely.

Ester Boserup's Sequence of Changing Land Use

Most scientists in fieldwork-based disciplines have a view of early land use that is much different from land-use modelers. Decades ago, the economist Ester Boserup published two influential books that synthesized the findings from a range of field studies: *The Conditions of Agricultural Growth: The Economics of Agrarian Change under Population Pressure* in 1965, and *Population and Technological Change: A Study of Long-Term Trends* in 1981. She proposed that a gradual shift in agricultural practices occurred over many millennia because of the invention and adaptation of new farming skills. In her view, those early farmers had indeed learned quite a few "new tricks" over all those thousands of years (Table 8-1). Decades later, Boserup's conceptual model remains the standard against which field-based efforts are compared.

During the earliest phase of long-fallow farming, agriculturalists shifted cultivation constantly from plot to plot (recall Chapter 4). To clear a new plot of land, farmers girdled trees to stop the flow of sap in the outer bark layers. After the trees died, they set fire to the dead debris during dry seasons, and then used dibble sticks to plant crop seeds in the ash-enriched soil. The ash also reduced the acidity typical of forest soils. After a few years of growing crops, soil fertility dropped and farmers moved to a new plot and repeated the sequence. Plots left lying fallow for several decades might be reoccupied, and new vegetation that had grown was again burned to enrich the soil (Figure 8-2).

TABLE 8-1   **The Boserup Sequence**

Long fallow $\rightarrow$ Short fallow $\rightarrow$ Annual crops $\rightarrow$ Double cropping

Earlier	Changes through time	Later
Low	Population density	High
Low	Labor required per acre	High
Low	Productivity per acre	High
High	Per capita acreage farmed	Low

FIGURE 8-2 Scrub vegetation and tree saplings growing on a cultivated plot several years after abandonment. *[Morley Read/Alamy.]*

This long-fallow phase of early agriculture used surprisingly large amounts of land. Each year's cultivated plot may have occupied perhaps 1 hectare (about 2 acres) per person, but the constant shifting from old to new plots left behind a large "footprint" long after each plot was abandoned (Figure 8-3). Because abandoned plots take decades to develop even semi-mature trees, only a small fraction of total reforestation would have occurred a few decades after abandonment. This sequence of leaving multiple footprints in effect boosted the cumulative area cleared per family at any time to many hectares.

Early farmers also cleared forests for other uses: pasture and hayfields to feed livestock; sites for houses and outbuildings; woodlots to supply timber for heating, cooking, homes, and outbuildings; and pathways and roads for travel. In many regions, early farmers also burned fields repeatedly to maintain clearings that attracted wild game to new growths of berries, nuts, and fresh grasses. At times, purposely set fires accidentally escaped into the wild. All these land uses added to the total clearance accomplished by early farmers. By several estimates, the per capita use of land by early farmers was large, perhaps 4–5 hectares (about 10 acres) per person. Early farmers had access to surprisingly

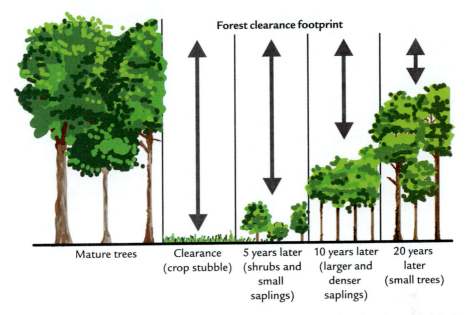

FIGURE 8-3 Shifting (long-fallow) cultivation clears new plots but leaves behind a footprint of incompletely reforested plots.

extensive land areas for a simple reason—population densities were very low and large areas could be farmed without encountering neighbors and causing conflicts.

Over many millennia, the reliable nutrition provided by agriculture led to population growth. As this happened, pioneering farmers acquired more neighbors, and claims on the available land began to increase. As farming families slowly became constrained to smaller holdings, they were forced to produce more food per hectare of land.

The gradual shift to smaller plots was made possible because new farming methods gradually shortened the fallow period from several decades to just a few years, and also lengthened the interval during which each plot could be actively cultivated. Key to these changes was a new emphasis on enriching soil quality, at first by mixing in grass and other available vegetable debris as fertilizer, and later by spreading animal and human manure. With these new techniques, farmers could gradually reduce the amount of land they used but keep total food productivity at the levels they needed.

Later, new methods that came into use 2,000 years ago in China and 500 years ago in Europe entirely eliminated the need to leave most land fallow. Farmers discovered the benefits of adding lime to reduce soil acidity, of rotating crops, and of growing legumes (peas, lentils, alfalfa, clover, and others) that extract nitrogen from the air and add it to the soil. With these new methods, farmers could now plant crops on the same plot of land every year, and in some areas they could plant two crops a year ("double cropping"). These innovations continued the long-term trend toward less cultivated land per person. By two or three centuries ago, most farmers in Eurasia cultivated less than 1 hectare per person, compared with 4–5 hectares several thousand years ago.

Along with improvements in soil treatment came technological innovations in the tools used to turn the soil and in the use of livestock as draft animals. As described in Chapter 4, over the course of many millennia farmers switched from poking crude dibble sticks into untilled ground, to using primitive ard plows that cut grooves in soil to make it easier to break up, to eventually manufacturing large V-shaped iron moldboard plows that were pulled by livestock and turned entire slabs of soil.

On first thought, it might seem that these labor-saving innovations would have enabled farmers to cultivate more land, which would be exactly counter to the previously described trend toward ever-smaller cultivated plots. But Boserup showed that these newer farming methods actually required ever-greater amounts of labor through time (we will ignore here the advent of mechanized farming during the 1800s).

Labor in the earliest stages of farming consisted of a few relatively simple chores: using flint axes to girdle trees and stop the flow of sap, waiting until dead debris piled up and setting it on fire, planting the seeds for new crops with sticks, keeping wild animals and birds away from the fields, and harvesting and storing the produce.

As new technologies allowed farmers to produce more food per hectare, the list of farm chores also grew. Crops still needed to be protected from wild animals and birds, and livestock still had to be guarded against predators and provided with summer pasture and winter fodder from summer crops. But now manure produced by the livestock had to be collected and spread on the fields. Weeds and insects had to be picked by hand to maximize crop yields. Boggy areas had to be drained to make them suitable for farming, and arid areas had to be irrigated so that crops didn't wither during dry summers. Drainage ditches and irrigation canals needed to be cleared of silt and woody debris to allow water to flow. Crops stored over the winter had to be protected against vermin and rot. As small family farmers today well know, the work is almost endless.

In summary, as new innovations such as better plows and increased use of livestock as draft animals emerged, their benefits were directed toward farming ever-smaller amounts of land more intensively, but using ever-greater labor. For example, rather than enabling farmers to till ever-larger areas of land, improved plows were used to till the same plots of soil more frequently (once or more a year) to counter the growth of weeds. In fact, farmers worked harder and harder to obtain their food from less and less land.

Historical Trends in Per Capita Land Use in Europe

Boserup did not attempt to estimate changes in per capita land use during the millennia when agriculture was gradually spreading across the continents, but historical data provide a means to do so and, thus, can be used to test the simplifying assumption of constant land use in many models.

The 2003 paper that introduced the early anthropogenic hypothesis emphasized the importance of the historical data point noted in Chapter 4: the Domesday Book survey of England and Wales ordered by William the Conqueror in 1086. Published in 1089, this survey counted a population of 1.5 million people and found that 85% of the arable land in England had been cleared (deforested).

The results of this survey contradict the proposed one-to-one link between population size and extent of forest clearance (Figure 8-4). Today, roughly 60 million people live in England and Wales, and over 80% of the arable land is cleared. For a one-to-one relationship between population and clearance, the 1.5 million people alive in 1089 should have cleared about 2.15% of the arable land. Yet the Domesday Book showed about 85% clearance, even more than today.

Because this survey was done almost 1,000 years ago, it might at first be thought to be wildly inaccurate. But Oliver Rackham, an exacting and meticulous botanist, decided to test its findings in every way imaginable. Among other things, he considered well-dated archeological remains of homes and villages, place names that constrain the founding of villages to particular eras, and the locations of modern woodlands, which usually match those in the Domesday survey closely. He concluded that the level of clearance in 1086 England was at least 85%, and perhaps as much as 90%: the Domesday survey was reliable. Because models based on a one-to-one link between land use and population would predict only about 2% clearance in 1089, the fundamental assumption of those models is clearly false.

In 2009, the climate modeler Jed Kaplan and colleagues further challenged the constant land-use assumption. They compiled historical

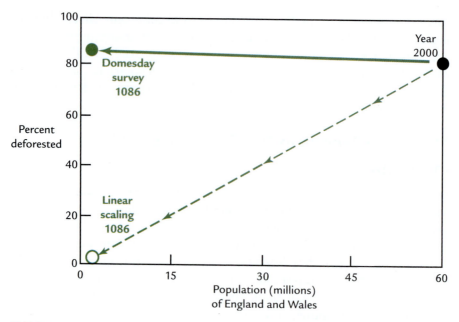

FIGURE 8-4 Estimated 2% land clearance in 1086 England and Wales based on backward projection of current clearance and population relationship, compared to 85% clearance reported by the Domesday Book.

records of forest clearance from France, Denmark, Sweden, Greece, and Ireland. Most of the records, spaced irregularly in time, came from the years 1000 to 1800, but a few went back to the start of the European historical era around 2,000 years ago. These historical data revealed surprisingly extensive early forest clearance long before the Industrial Era, consistent with the Domesday Book.

By plotting percent of forest clearance against population density, Kaplan and colleagues found that forest clearance does not track population in a one-to-one way. Instead, clearance accelerates quickly at low population densities of about 10 people per square kilometer, and is 80 to 90% complete by the time the population density reaches intermediate levels of 100 people per square kilometer (Figure 8-5). Additional increases to higher modern population densities of 1,000 people per square kilometer have little or no effect on clearance because, by this time, most of the forests have already been cleared.

Kaplan used historical data to estimate pre-industrial forest clearance across all of Europe. His reconstructions differ strikingly from those based on the assumption of constant per capita land use. For example, a frequently cited land-use reconstruction based on nearly

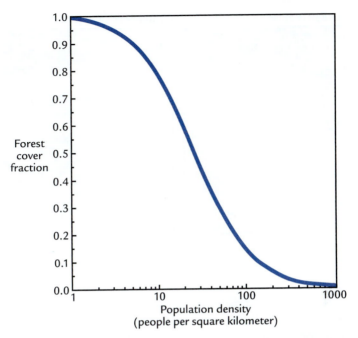

FIGURE 8-5 Relationship between forest clearance and population density based on historical evidence from Europe. *[Adapted from J. O. Kaplan, K. M. Krumhardt, E. C. Ellis, W. F. Ruddiman, C. Lemmen, and K. K. Goldewijk, "Holocene Carbon Emissions as a Result of Anthropogenic Land Cover Change," The Holocene 25 (2011): 775–791, doi:10.1177/0959683610386983.]*

constant land use through time (labeled "HYDE" in Figure 8-6), shows most of Europe still largely forested in 1800, but Kaplan's simulation based on historical data shows Europe largely deforested, in agreement with conclusions reached by many archeologists, paleoecologists, and others who have conducted field studies of land use (recall Chapter 4). By 1800, the only large remaining area of uncut forest in Kaplan's reconstruction, aside from a few regions of higher alpine terrain, was western (European) Russia.

Kaplan's reconstruction of land use built on earlier studies of the historian Alexander Mather, who found that much of Europe was deforested many centuries ago and had actually begun *re*foresting during the 1800s and 1900s, as population levels soared toward their modern levels. Mather called this surprising reversal in the deforestation trend the **forest transition**, a transition to an entirely new relationship between population and clearance.

Clearance at 1800 (Kaplan)   Clearance at 1800 (HYDE)

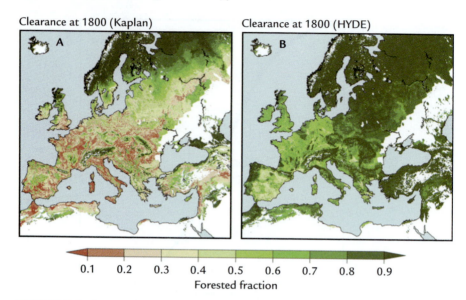

FIGURE 8-6 Contrasting estimates of European forest cover in 1800. *[Adapted from J. O. Kaplan, K. M. Krumhardt, and N. Zimmerman, "The Prehistorical and Preindustrial Deforestation of Europe,"* Quaternary Science Reviews *28 (2009), 3016–3034.]*

For Europe as a whole, the constant land-use assumption fails to capture the enormous extent of early clearance. The problem with modeling efforts based on this assumption is that such models overlook a host of studies in field sciences summarizing evidence about how early people farmed the land—such as the evidence summarized in Chapter 4.

Historical Trends in Per Capita Land Use in China

As noted in Chapter 5, John Buck compiled available land-use trends for the entire agricultural area of east-central China over the last 2,000 years. The area surveyed included dryland crops like millet, soybeans, and wheat in the north, and irrigated rice in the south. The historical trend showed a decrease in per capita area cultivated from 0.6–0.7 hectares per person nearly 2,000 years ago to roughly 0.15–0.2 hectares by the early 1800s (Figure 8-7).

Buck's research covers cultivated plots but omits pastures, hayfields, houses, roads, woodlots, and fallow land. More importantly, it omits regions covered by water as well as roughly 60% of the land in east-central China that is unsuitable for agriculture because of steep and rocky terrain. Most of these hillsides and rocky slopes were cleared of

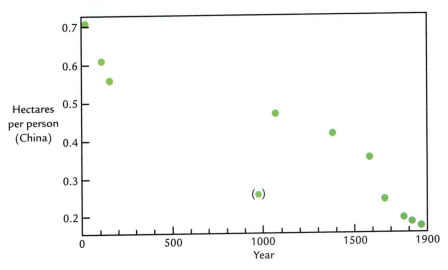

FIGURE 8-7 Cultivated land per person in China from 2,000 years ago to the 1800s. (The anomalously low cultivation value at the year 976 followed shortly after a major war that left much of the land temporarily unusable.)[*Based on data in J. L. Buck,* Land Utilization in China *(Oxford: Oxford University Press, 1937); and K. Chao,* Man and Land in Chinese History: An Economic Analysis *(Stanford: Stanford University Press, 1986).]*

trees long ago by already-dense populations of people seeking firewood for cold winters and lumber for farm structures. Including these additional areas would increase the per capita calculation of forest clearance shown in Figure 8-7 by a factor of about 2.5, so that the actual land-use trend would have fallen from roughly 1.6 hectares per person 2,000 years ago to about 0.4 hectares in the 1800s. This range in per capita land use is similar to that calculated from the historical surveys in Europe by Jed Kaplan and colleagues. The evidence from these two historical studies confirms the decrease in per capita forest clearance during the historical era implicit in Ester Boserup's efforts.

In 2010, Jed Kaplan and colleagues extended their simulations of forest clearance to a global scale, using the European historical evidence on population density and per capita clearance as a start, but adjusting for the longer growing season and higher productivity typical of tropical latitudes. Their method captured the greater clearance in and near cities and towns with high populations and the lesser clearance in the countryside.

Their method simulated major deforestation by 2,000 years ago in Europe, China, India, the eastern Mediterranean, Peru, and Mexico—all areas of relatively dense populations (Figure 8-8A). Early farming populations in these regions had already caused large amounts of deforestation

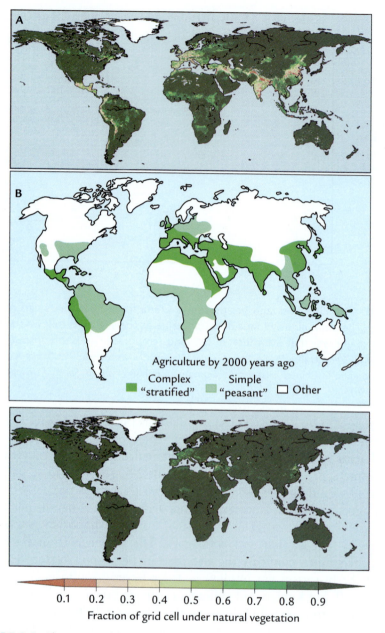

FIGURE 8-8 Clearance of forest and other vegetation 2,000 years ago based on large per capita clearance (A) and small, constant per capita clearance (C), compared to map of agriculture based on archeological information (B).

[Both (A) and (C) adapted from J. O. Kaplan, K. M. Krumhardt, E. C. Ellis, W. F. Ruddiman, C. Lemmen, and K. K. Goldewijk, "Holocene Carbon Emissions as a Result of Anthropogenic Land Cover Change," The Holocene 25 (2011): 775–791, doi:10.1177/0959683610386983. (B) based on J. W. Lewthwaite and A. Sherritt, "Chronological Atlas," in Cambridge Encyclopedia of Archaeology (Cambridge: Cambridge University Press, 1986).]

millennia ago, long before the Industrial Revolution. Significant clearance in these areas of high population density, early empires, and intensive agriculture matches previous archeology-based estimates (Figure 8-8B) such as the efforts reviewed in Part 2. In contrast, simulations based on models that rely on the assumption of small per capita land use simulate almost no clearance in these regions, or anywhere else (Figure 8-8C).

Changing Land Use for Rice Irrigation

So far we have focused on forest clearance, which generates CO_2 emissions. But what about farming activities that generate CH_4? Did long-term trends in per capita land use also occur in the areas where those farming practices were used? Although the rise in methane concentrations that began 5,000 years ago looks somewhat more similar to the increase in population than does the CO_2 trend (see Figure 8-1), the CH_4 trend again rose faster and earlier than population. If humans were the cause of the CH_4 increase, factors like early rice farming and livestock tending must have emitted considerably more CH_4 per person long ago than they did later.

Historical evidence

Growing rice is a highly intensive kind of farming: it produces a large amount of food per hectare of land, but it also requires enormous amounts of human labor, even today (Figure 8-9). The Tai Lake region in the lower reaches of the Yangtze River valley (recall Chapter 5) has been a center of irrigated rice cultivation for thousands of years, and its historical record includes a dozen surveys of land used per person over the last 1,000 years (Figure 8-10). Near the year 1000, per capita area of land farmed in rice paddies averaged about 0.7 hectares, but by the 1700s, per capita use had dropped to approximately 0.17 hectares, a four-fold decrease in less than 1,000 years.

As was the case in dryland farming in Europe and northern China, this trend of decreasing per capita land use resulted from improved farming practices: increased use of manure and ash from fires; more intensive weeding and pest control (with the increased labor available from the rising populations); and planting crops every year, or twice a year. Water management also became more sophisticated: chain and bucket pumps, tanks, and reservoirs lifted and stored water; canals and sluiceways moved it to the fields; and paddies were leveled to distribute the water more evenly on the fields.

These innovations led to greater food production per acre, food that was badly needed to offset the huge increase of population compared to the much smaller increase in amount of land farmed.

FIGURE 8-9 Chinese farmers working in flooded rice paddies. *[dbimages/Alamy.]*

By around the year 1400, every piece of arable land in the Tai Lake region was in use, and the small additional increase in per capita land use from 1400 to 1550 resulted from the conversion of brackish low-lying areas to rice paddies.

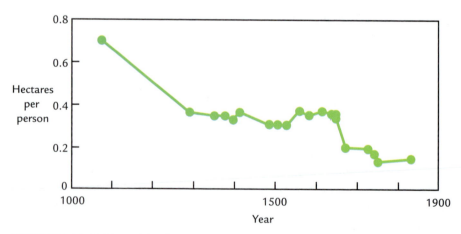

FIGURE 8-10 Cultivated land per person in Tai Lake, China, during the last millennium. *[Adapted from E. C. Ellis and S. M. Wang, "Sustainable Traditional Agriculture in the Tai Lake Region of China,"* Agriculture, Ecosystems, & Environment 61 (1997): 177–193.]

The declining historical trends in land use in this region help to explain why CH_4 emissions (and atmospheric concentrations) rose earlier than the growth in population. Early farmers needed to irrigate far more land per capita than later ones, and thus emitted more methane per person than they did later.

Another (less well-documented) trend in recent centuries in densely populated areas like eastern China was a gradual shift to fewer livestock per person. Very early on, when populations were small and land was readily available, there were few limitations on keeping large herds. Farmers could choose to use their land for some combination of growing crops and tending livestock that needed summer pasture and hayfields to provide fodder for winter. In these earlier times, farming families also obtained meat and milk from their livestock.

Later, as populations grew and land became much more crowded, farmers had to survive on smaller and smaller plots. This change forced them to choose between keeping livestock (which take up a great deal of land relative to their nutritional return) and growing crops (which provide more nutrition per hectare than livestock). Under this pressure, farmers gradually began to devote more of their land to rice farming, eventually keeping only enough livestock to pull plows and provide manure, milk, and cheese, but not slaughtering their animals for meat. Groups of farming families in villages also began to share oxen. This slow trend toward fewer livestock per farming family further reduced per capita emissions of methane.

Archeological Evidence

Archeological data also reveal greater earlier per capita land use for rice irrigation. As shown in Chapter 5, Dorian Fuller and colleagues mapped the spread of irrigated rice farming across all of southern Asia (recall Figure 5-14). They then estimated that rice farming within each region would subsequently have increased with the square root of the population density, the relationship observed in modern agriculture in the region. By combining the first arrival of irrigated rice in each area and the later increase in density, Fuller and colleagues estimated the increase in total irrigated area across southern Asia from 5,000 to 1,000 years ago (Figure 8-11).

According to Fuller and colleagues, farmers by 1,000 years ago were irrigating an estimated 38% of the total amount irrigated today, yet the population in southern Asia at that time was only 6% of the current level. The ratio between these two numbers (38% and 6%) indicates that early rice farmers had brought under irrigation more than

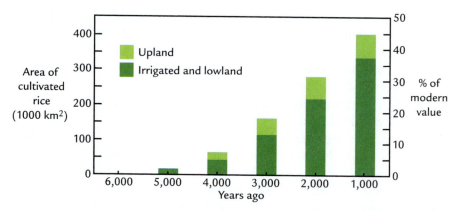

FIGURE 8-11 Estimated increase in land irrigated for rice farming from 5,000 to 1,000 years ago as a percent of modern value. *[Adapted from D. Q. Fuller, J. Van Etten, K. Manning , C. Castillo, E. Kingwell-Banham, A. Weisskopf, L. Qin, Y.-I. Sato, and R. Hijmans. "The Contribution of Rice Agriculture and Livestock Pastoralism to Prehistoric Methane Levels: An Archeological Assessment," The Holocene 25 (2011): 743–759, doi:10.1177 /0959683611398052.]*

six times the amount of land that would be expected based on their small population numbers. Fuller's estimate shows that the early increase in extent of irrigated rice agriculture far preceded the major rise in population during the last 5,000 years, again supporting the Boserup sequence.

The Significance of Decreasing Per Capita Land Use

If per capita land use is assumed to have remained constant for thousands of years, the total amount of land cleared (and carbon emitted into the atmosphere) would inevitably track the exponential rise in human population. Most of the forest clearance and associated carbon emissions would then have occurred during the last few centuries (Figure 8-12A).

But if per capita land use was higher thousands of years ago and then fell over time, as indicated by the wide range of historical and archeological data summarized above, the amount of clearance millennia ago would have been greater than implied by the small populations of the time, while in recent centuries it would not have risen as fast as the surging population trend would suggest. If the exponential increase in recent population were exactly mirrored by an opposing exponential decrease in per capita land use, total clearance through time would have followed a nearly linear increase (see Figure 8-12B). This trend would have substantially reduced the apparent offset between the greenhouse-gas and population trends in Figure 8-1.

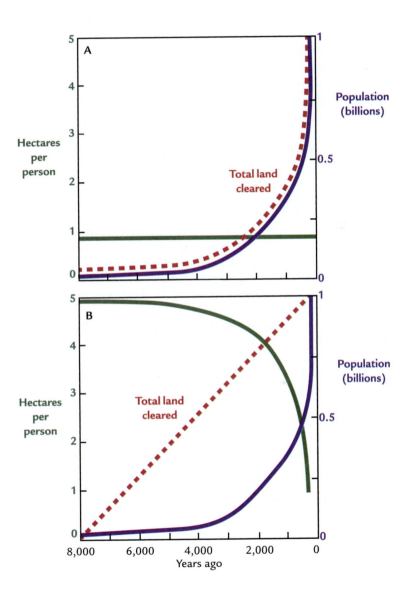

FIGURE 8-12 Trend in increasing cumulative amount of land cleared through time differs considerably for (A) constant amount of clearance per person versus (B) decreasing per capita clearance (as observed in historical records).

Summary

A wide range of historical evidence supports large per capita clearance and land use millennia ago. Although reliable records of early land use are somewhat fragmentary, they clearly agree that land use millennia ago was higher than in recent (pre-industrial) centuries. Historical evidence shows a four-fold decrease in per capita farm size over the last 2,000 years in both Europe and China. Estimates based on archeological data show a similar trend prior to the historical era in southern Asia. This historical and archeological evidence invalidates model simulations based on assuming small, constant per capita clearance and cultivation through pre-industrial time. The heavily forested, nearly pristine continents previously simulated as having persisted into late pre-industrial time are not consistent with this historical and archeological evidence.

Additional Resources

Boserup, E. *The Conditions of Agricultural Growth: The Economics of Agrarian Change Under Population Pressure*. London: Allen and Unwin, 1965.

Boserup, E. *Population and Technological Change: A Study of Long-Term Trends*. Chicago: University of Chicago Press, 1981.

Buck, J. L. *Land Utilization in China*. Shanghai: Commercial Press, 1937.

Chao, K. *Man and Land in Chinese History: An Economic Analysis*. Stanford: Stanford University Press, 1986.

Ellis, E. C., and S. M. Wang. "Sustainable Traditional Agriculture in the Tai Lake Region of China." *Agriculture, Ecosystems, & Environment* 61 (1997): 177–193.

Fuller, D. Q., J. Van Etten, K. Manning, C. Castillo, E. Kingwell-Banham, A. Weisskopf, L. Qin, Y.-I. Sato, and R. Hijmans. "The Contribution of Rice Agriculture and Livestock Pastoralism to Prehistoric Methane Levels: An Archeological Assessment." *The Holocene* 25 (2011): 743–759. doi:10.1177/0959683611398052.

Goldewijk, K. K. "Estimating Global Land Use Change Over the Last 300 Years: The HYDE Database." *Global Biogeochemical Cycles* 15 (2001): 417–433.

Grigg, D. "Ester Boserup's Theory of Agrarian Change: A Critical Review." *Progress in Human Geography* 3 (1979): 64–84. doi: 10.1177/030913257900300103.

Kaplan, J. O., K. M. Krumhardt, and N. Zimmerman. "The Prehistorical and Preindustrial Deforestation of Europe." *Quaternary Science Reviews* 28 (2009), 3016–3034.

Kaplan, J. O., K. M. Krumhardt, E. C. Ellis, W. F. Ruddiman, C. Lemmen, and K. K. Goldewijk. "Holocene Carbon Emissions as a Result of Anthropogenic Land Cover Change." *The Holocene* 25 (2011): 775–791. doi:10.1177/0959683610386983.

Kirch. P. V. "Archaeology and Global Change: The Holocene Record." *Annual Review of Environment and Resources* 30 (2005): 409–440, doi:10.1146 /annurev.energy.29.10243.140700.

Ramankutty, N., and J. A. Foley. "Estimating Historical Changes in Global Land Cover: Croplands from 1700 to 1992." *Global Biogeochemical Cycles* 13 (1999): 997–1027.

Ruddiman, W. F., and E. C. Ellis. "Effect of Per-capita Land-use Changes on Holocene Forest Clearance and CO_2 Emissions." *Quaternary Science Reviews* 28 (2009): 3011–3015. doi:10.1016/j.quascirev.2009.05.022.

CHAPTER 9

HOW SHOULD INTERGLACIAL GAS TRENDS BE COMPARED?

Greenhouse-gas concentrations rose in the later part of the current interglaciation but fell during the equivalent parts of prior interglaciations (recall Part 1, Chapters 2 and 3). When the debate over the origins of these trends began, ice-core drilling had fully penetrated only the three interglaciations (named stages 5, 7, and 9) prior to the current one (named stage 1). Ice layers spanning previous interglaciations lay within reach of drilling at other locations, but had not yet been recovered. Subsequent drilling by the European Project for Ice Coring in Antarctica (EPICA) has now penetrated several additional interglaciations (through stage 19) at a site called Dome C. This more recent drilling brings the number of previous interglaciations available for examination to seven.

As this deeper ice was being recovered, another major challenge to the early anthropogenic hypothesis arose—the issue of how the early-interglacial CO_2 and CH_4 trends in previous interglaciations should be aligned against those in the current one. Obtaining the correct (or at least the best possible) alignment is necessary in order to confirm or refute the conclusions tentatively reached in Chapters 2 and 3. Did gas trends in these newly recovered interglaciations fall as they did in the three interglaciations examined in Chapters 2 and 3, or did they instead increase, the trend observed during the current interglaciation?

This question is important, because previous interglaciations record the natural behavior of the climate system. Our human ancestors were roaming Earth during those much earlier times, but they were far too small in numbers (probably a few million at most), and they also lacked the technological skills necessary to influence global gas emissions in a major way. Consequently, if the past interglacial gas trends actually rose, that would suggest that the upward trends in this interglaciation could have been natural; but if they fell, that would indicate that this interglaciation is different (not natural), presumably because of some new influence (human activities).

Two different methods have been used to align gas trends during the various interglaciations. The method discussed in Part 1 relies on summer solar radiation (insolation). An alternative method involves aligning the records based on the deglacial intervals just prior to the various interglaciations.

Aligning Gas Trends to Orbital Changes: The Insolation-Alignment Method

The insolation-alignment method based on summer solar radiation (insolation) trends involves two steps. First, ages for the entire sequence of ice layers are estimated using physical **ice flow models** of how the snow deposited on top of the ice sheets slowly consolidates into ice and then flows downward and outward from the center of the ice sheet under its own weight. The EPICA group applied this method to estimate the ages of the entire 800,000-year-old ice sequence drilled at the Dome C site, and this chapter adopts their widely used time scale (referred to as EDC3).

The second step starts with the precisely known ages of past insolation trends based on calculations of Earth's orbit from astronomy. Because of Earth's orbital cycles, similar maximum and minimum values of summer insolation recur early in all interglaciations at accurately known times. The calculated ages of these maxima and minima based on astronomy are used as times for aligning the EPICA-estimated ages for the sequence of ice layers. In this way, gas trends in the ice during the various interglaciations are aligned to insolation trends.

With the additional older interglaciations now retrieved by drilling at Dome C, this method can be further tested, again using insolation trends as the basis for the alignment (Figure 9-1). The ages of all of the interglaciations examined are shown in Table 9-1. Note that stages 3 and 13 are omitted from this analysis because they didn't reach full interglacial amplitudes.

Based on the insolation-alignment method, the CH_4 concentrations that rose during the last 5,000 years of the current (stage 1) interglaciation fell during the equivalent interval in all seven previous interglaciations (Figure 9-2). These downward trends in the past

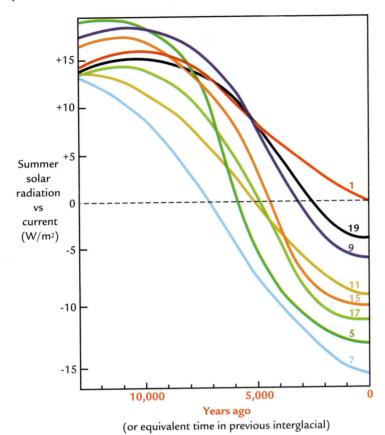

FIGURE 9-1 Changes in summer solar radiation during the current interglaciation, compared to the seven preceding ones. *[Adapted from A. Berger, "Long-term Variations of Caloric Insolation Resulting from Earth's Orbital Elements,"* Quaternary Research 9 (1978): 139–167.]

TABLE 9-1 Ages of Interglaciations

Name of interglaciation	Age (years ago)
Stage 1	Present day
Stage 5	120,000
Stage 7	240,000
Stage 9	330,000
Stage 11	405,000
Stage 15	610,000
Stage 17	695,000
Stage 19	785,000

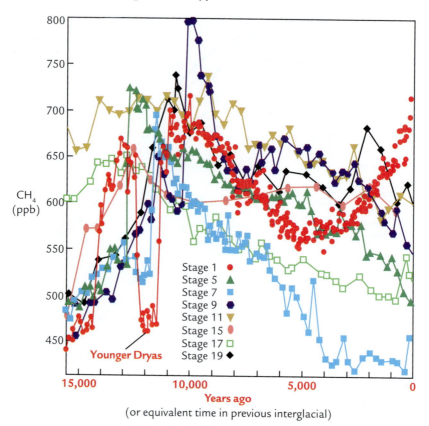

CH$_4$
(ppb)

Stage 1 ●
Stage 5 ▲
Stage 7 ■
Stage 9 ⬣
Stage 11 ▼
Stage 15 ●
Stage 17 □
Stage 19 ◆

Younger Dryas

15,000   **10,000**   **5,000**   **0**

Years ago
(or equivalent time in previous interglacial)

FIGURE 9-2 Trends in atmospheric methane concentrations early in eight interglaciations based on the insolation-alignment method. *[CH$_4$ concentrations are from EPICA Community Members, "Eight Glacial Cycles from an Antarctic Ice Core,"* Nature 429 *(2004): 623–628.]*

unanimously identify the rising trend during the present-day inter-glaciation as anomalous (Table 9-2).

The verdict on the CO$_2$ concentrations is similar, but not unani-mous (Figure 9-3, Table 9-3). During the times equivalent to the last 7,000 years in stage 1, six of the seven CO$_2$ trends during previous interglaciations fell—some by large amounts, and some just a little. Only one previous interglaciation (stage 15) shows a clear rising trend, and the increase (about 8 ppm) is much smaller than the rise in the current interglaciation (roughly 22 ppm).

The overall message from the insolation-alignment method is clear: thirteen of the fourteen CH$_4$ and CO$_2$ trends head downward, completely unlike the increases late in the current interglaciation. These comparisons

TABLE 9-2 CH$_4$ Trends During Previous Interglaciations: Insolation-Alignment Method

Natural explanations predict **CH$_4$ increases (like that in the current interglaciation).**

Anthropogenic explanation predicts CH$_4$ decreases (unlike the increase in the current interglaciation).

Interglaciation	CH$_4$ trend
Stage 5	decrease
Stage 7	decrease
Stage 9	decrease
Stage 11	decrease
Stage 15	decrease
Stage 17	decrease
Stage 19	[decrease]

[ ] Small trend

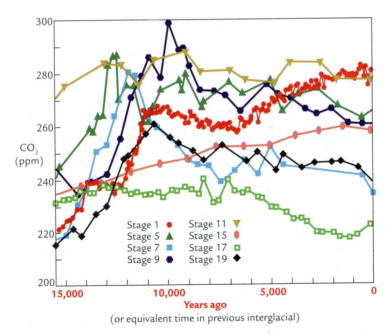

FIGURE 9-3 Trends in atmospheric carbon dioxide concentrations early in eight interglaciations based on the insolation-alignment method. *[CO$_2$ concentrations are from EPICA Community Members, "Eight Glacial Cycles from an Antarctic Ice Core," Nature 429 (2004): 623–628.]*

TABLE 9-3 CO_2 **Trends During Previous Interglaciations: Insolation-Alignment Method**

Natural explanations predict CO_2 increases (like that in the current interglaciation).

Anthropogenic explanation predicts CH_4 decreases (unlike the increase in the current interglaciation).

Interglaciation	CO_2 trend
Stage 5	[decrease]
Stage 7	decrease
Stage 9	decrease
Stage 11	[decrease]
Stage 15	increase*
Stage 17	decrease
Stage 19	decrease

[ ] Small trend
*Anomalous ice-volume trend

strongly contradict natural explanations for the gas increases during the last several thousand years and favor an anthropogenic one.

Aligning Gas Trends to the Start of Previous Deglaciations: The Deglacial-Alignment Method

The question posed at the beginning of this chapter remains. Would a different way of aligning greenhouse-gas trends produce the same result as the insolation alignments?

The EPICA group in 2004 and a paper by the climate scientists Wallace Broecker and Thomas Stocker in 2006 used an alternative technique to align the current interglaciation against one of the previous interglaciations (stage 11, dating to about 400,000 years ago). The two groups aligned the rapid CO_2 rise that occurred early in the deglaciation prior to interglacial stage 11 against the rapid CO_2 rise that occurred early in the deglaciation prior to the current interglaciation. Based on this alignment, they then compared the rest of the CO_2 and CH_4 trends in the stage 1 and stage 11 interglaciations that followed.

But this alignment method is not an independent way of comparing gas trends in all seven previous interglaciations against those in the present one, because the gas trends themselves are used to create the alignment that is to be tested. Instead, we need an independent basis for aligning the early parts of the various deglaciations.

Fortunately, such a technique was introduced in Chapter 1—the oxygen-isotope index that measures how much ^{16}O-rich ocean water was evaporated from the ocean and stored in ice sheets, leaving relatively ^{18}O-rich water in the ocean. As the ice sheets melted during deglaciations and returned ^{16}O-rich water to the ocean, the oxygen-isotope index in the water shifted to lighter (^{16}O-rich) values that were recorded in the shells of marine organisms that accumulated in the permanent sedimentary record on the seafloor.

Figure 9-4 shows the oxygen-isotope records from all eight deglaciations (the most recent and the seven earlier ones) aligned on the levels (times) when ice volume began to decrease, and plotted against the time elapsed until most ice had finished melting (length of deglaciation). Four of the previous interglaciations (stages 5, 7, 9, and 19) had relatively fast deglaciations (lasting about 10,000 years), similar to the one leading into

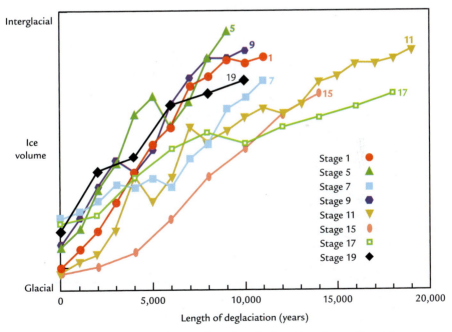

FIGURE 9-4 Melting trends for northern hemisphere ice sheets for eight deglaciations. *[Adapted from marine oxygen-isotope of L. E. Lisiecki and M. E. Raymo, "A Plio-Pleistocene Stack of 57 Globally Distributed Benthic $\delta^{18}O$ Records,"* Paleoceanography *20 (2005): PA1003, doi:10.1029/2004PA001071.]*

the current interglaciation. Three other interglaciations (stages 11, 15, and 17) were preceded by slower deglaciations that lasted from 14,000 years to as much as 19,000 years.

The deglaciation trends shown in Figure 9-4 can be used to align the gas trends during all of the interglaciations. The beginnings of all eight deglaciations are set to a common reference age (marked "0" in Figure 9-4), and the subsequent gas concentrations are plotted in "elapsed time" after those reference points.

The greenhouse-gas trends that result from using this deglacial-alignment method show a range of patterns compared to the steady rises in CH_4 and CO_2 during the last several thousand years of the current interglaciation (Figure 9-5). The four interglaciations (5, 7, 9,

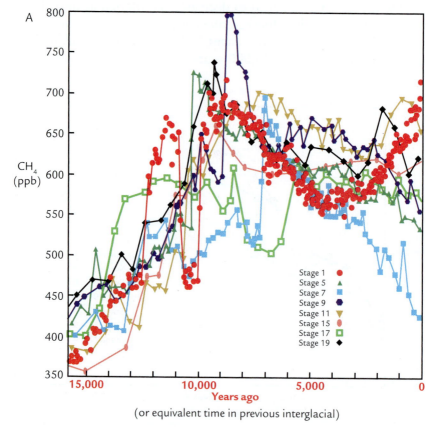

FIGURE 9-5 Trends in atmospheric methane and carbon dioxide concentrations in eight interglaciations using the deglacial-alignment method. [CO_2 and CH_4 concentrations are from EPICA Community Members, "Eight Glacial Cycles from an Antarctic Ice Core," Nature 429 (2004): 623–628.]

and 19) that occurred subsequent to the shorter (10,000-year) deglaciations all show downward gas trends during the timespan of the stage 1 greenhouse-gas rises, while the three interglaciations (11, 15, and 17) that followed the longer deglaciations show varying trends. For both CO_2 and CH_4, one of the latter three interglaciations shows a small downward gas trend, one shows no obvious or persistent trend, and one follows an upward trend (Table 9-4).

Using the deglacial-alignment method, five of the seven test cases show downward gas trends like those from the insolation-alignment method. This finding adds further support to the interpretation that the upward CO_2 and CH_4 trends during the current interglaciation are anomalous, rather than natural. On balance, the deglacial-alignment method favors the anthropogenic explanation, aside from the disagreements noted in Table 9-4. To explore the reasons for these disagreements, we need to assess how well-justified the two alignment methods are.

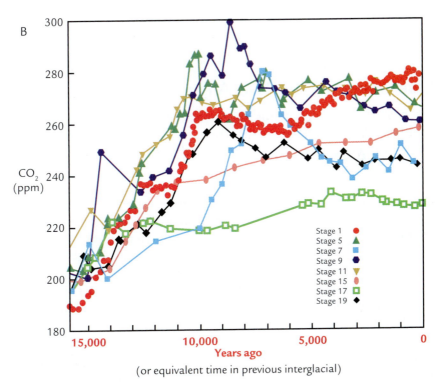

FIGURE 9-5 (Continued)

TABLE 9-4 CO_2 and CH_4 Trends During Previous Interglaciations: Deglacial-Alignment Method

Natural explanations predict CH_4 and CO_2 increases (like that in the current interglaciation).

Anthropogenic explanation predicts CH_4 and CO_2 decreases (unlike the increase in the current interglaciation).

Interglaciation	CH_4 Trend	CO_2 Trend
Stage 5	decrease	[decrease]
Stage 7	decrease	decrease
Stage 9	decrease	decrease
Stage 11	decr/incr	[decrease]
Stage 15	—	increase*
Stage 17	[decrease]	—
Stage 19	[decrease]	[decrease]

[ ] Small trend
*Anomalous ice-volume trend
—No net overall trend

Assessment of the Two Alignment Methods

The primary justification for the insolation-alignment method comes from basic knowledge about the causes of atmospheric methane variations (recall Chapter 2). Geochemists analyzing ice-core methane trends have long attributed a significant part of the CH_4 signal to the effects of the strong northern monsoons stretching from Africa across India to China (see Chapter 2, Figure 2-4). In these regions, monsoons driven by summer insolation alternately filled methane-emitting wetlands or left them dry, thus contributing significantly to natural variations in methane concentration in the atmosphere. Based on this link, monsoons driven by orbital insolation in the northern tropics have been widely regarded as a primary (but not the sole) influence on past CH_4 changes.

This monsoon explanation works well for the start of the current interglaciation. The summer insolation maximum 11,000 to 10,000 years ago matches the most recent CH_4 maximum, which has been very accurately dated in the Greenland ice sheet by counting annually deposited layers (Figure 9-6). The monsoon explanation is also supported by well-dated cave deposits from China spanning three previous interglaciations

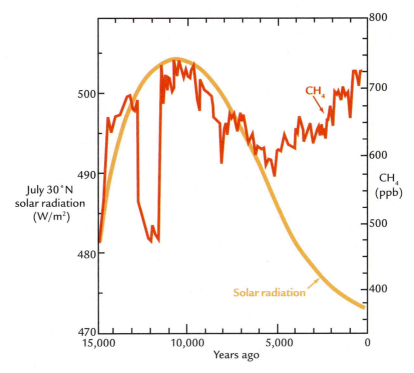

FIGURE 9-6 Trends in 30°N summer insolation (from orbital calculations) and in CH_4 concentrations (from annually layered ice) for the last 15,000 years. *[Summer insolation from A. Berger, "Long-term Variations of Caloric Insolation Resulting from Earth's Orbital Elements," Quaternary Research 9 (1978): 139–167; methane concentrations from T. Blunier et al., "Variations in Atmospheric Methane Concentration During the Holocene Epoch," Nature 374 (1995): 46–49.]*

during the last 350,000 years. These cave records show changes in oxygen-isotope composition that were thought to have been caused by variations in the Asian summer monsoon in response to insolation changes at the orbital precession cycle (every 23,000 years). The connection between summer insolation and the north-tropical monsoon response is very clear.

Fluctuations in the northern monsoons at the 23,000-year cycle are not the only factor that has affected atmospheric CH_4 concentrations (see Chapter 2). Changes in Arctic wetlands are a second influence, probably at least in part under the impact of the strong 41,000-year orbital insolation signal at high latitudes. Still, the clear links between insolation, northern monsoon responses, and methane concentrations provide a good justification for aligning the CH_4 trends in previous interglaciations to the orbital time scale of northern hemisphere insolation.

In addition, both CO_2 and CH_4 gases are trapped in the same bubbles of air in the ice. As a result, a time scale justified by a link between insolation and CH_4 can be directly transferred to the CO_2 variations recorded in the same air bubbles.

Interglacial stage 15, with its slowly rising CO_2 trend, is an obvious outlier among the 14 test-case comparisons of the insolation-alignment method (see Tables 9-2 and 9-3). But stage 15 is also anomalous in another respect. The ^{18}O trend rose throughout that interval, suggesting that ice sheets were melting throughout the early part of the stage 15 interglaciation, rather than reaching a stable ice-volume minimum or starting to grow again. Because CO_2 levels and ice volume have been tightly linked for hundreds of thousands of years (see Chapter 3, Figure 3-2), this anomalous trend in ice volume might explain the different CO_2 response.

During the current interglaciation, the last ice (in Canada) had finished melting by 7,000 years ago, but apparently some northern hemisphere ice was still melting during the equivalent part of interglacial stage 15. This contrast suggests that interglacial stage 15 may not be valid for comparison to the current interglaciation. In any case, the six other test cases for CO_2 show no persistent upward trend during the early parts of the interglaciations.

The alternative deglacial-alignment method, which anchors comparisons between stages on the start of preceding deglaciations, carries an implicit assumption. If every deglaciation proceeded in more or less the same way, then this method should align the subsequent interglaciations in a consistent manner. But Figure 9-4 shows a problem with this assumption: the lengths of the eight deglaciations varied from about 10,000 years to just under 20,000 years. This wide range of variation in length must have had an effect on the relative alignments of the subsequent interglaciations.

The four previous interglaciations that were preceded by shorter deglaciations (stages 5, 7, 9, and 19) all show falling gas concentrations during the interval when gas levels were rising during the current interglaciation. Because all five of these deglaciations had similar (roughly 10,000-year) lengths, the deglacial-alignment method did not produce significant offsets in the relative timing of the subsequent interglaciations. The downward gas trends during these four previous interglaciations agree with those determined from the insolation-alignment method, and support the early anthropogenic hypothesis.

In contrast, the three interglaciations that were preceded by longer deglaciations of varying length (stages 11, 15, and 17) show an inconsistent mixture of gas trends during the times equivalent to the last several thousand years of the current interglaciation (see Table 9-4).

These varying trends suggest that the differing lengths of these longer deglaciations may complicate or perhaps invalidate this method of aligning the subsequent interglaciations (and their gas trends).

Interglacial Stage 11: A Special Case

Particular attention has been focused on interglacial stage 11, the one preceded by the longest (almost 20,000-year) deglaciation (see Figure 9-4). The reason for this attention is that during this time period, around 400,000 years ago, the eccentricity of Earth's orbit was very low, similar to the configuration in the current interglaciation. This similarity makes stage 11 a possible insolation analog to the present day.

Another reason for the interest in stage 11 is that both the EPICA group and Broecker and Stocker interpreted it as having lasted about 26,000 years. By aligning the start of this long interglaciation against the start of the current one (which at this point has lasted less than 10,000 years), they estimated that the current interglaciation should last for another 16,000 years or more (ignoring modern-day anthropogenic interference in the climate system). If their conclusion is correct, it would obviously invalidate the early anthropogenic hypothesis, which claims that CO_2 and CH_4 trends should be dropping by now, and climate should be cooling.

Subsequent evidence has refuted the central arguments of this challenge. First, a closer examination of the evidence reveals that the fully developed stage 11 interglaciation only lasted for 10,000 years or less, not 26,000 years. Evidence for this conclusion comes from the marine oxygen-isotope index and from measurements of ancient sea levels in the Red Sea, both of which are tied to global ice volume (Figure 9-7). Both the EPICA and Broecker-Stocker studies had interpreted a long interval late in the deglaciation that preceded interglacial stage 11 as being a fully developed interglaciation, when in fact some glacial ice was still left on northern hemisphere land masses (Canada and possibly Eurasia) at that time. The full peak stage 11 interglaciation, with sea level at a position similar to today, only lasted from about 410,000 to about 400,000 years ago, roughly the same length of time as the current interglaciation (to date). This finding invalidates the conclusion that the current interglaciation still has 16,000 years left to run. This finding implies that the current interglaciation should be ending, if not already at an end.

The second new finding is that the deglacial-alignment method does not align interglacial stages 11 and 1 correctly (see Figure 9-7). Both deglaciations began in a similar way with rapid ice melting, but later

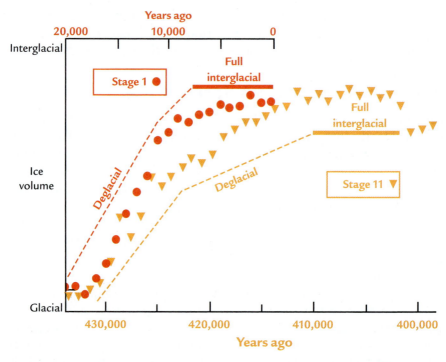

FIGURE 9-7 Comparison of global ice-volume signals during interglacial stage 11 with the current interglaciation (stage 1) based on the deglacial-alignment method. *[Ice volume based on $^{16}O/^{18}O$ record from L. E. Lisiecki and M. E. Raymo, "A Plio-Pleistocene Stack of 57 Globally Distributed Benthic $\delta^{18}O$ Records," Paleoceanography 20 (2005): PA1003, doi:10.1029 /2004PA001071.]*

the similarity broke down. The most recent deglaciation continued at a rapid rate and ended relatively quickly (within about 10,000 years), but the last part of the deglaciation that preceded stage 11 was much slower and lasted nearly twice as long (almost 20,000 years).

Using the deglacial-alignment method, the differing lengths of these two deglaciations result in such a large misalignment between the two full interglaciations that they don't overlap (see Figure 9-7). The current interglaciation falls entirely within the long, slow end of the deglaciation preceding stage 11, and it ends before the full stage 11 interglaciation begins. Sea-level records from the Red Sea tell much the same story. Examined closely, the deglacial-alignment method does not work as intended, and it is not a valid way of comparing the stage 11 and stage 1 interglaciations.

In contrast, the insolation-alignment method indicates that CH_4 and CO_2 greenhouse-gas concentrations would have been falling during the time equivalent to the present day, causing climate to cool (see Figures 9-2 and 9-3). The oxygen-isotope and Red Sea findings also indicate that small ice sheets would have been growing somewhere on the northern hemisphere continents, causing sea levels to fall. Rather than contradicting the early anthropogenic hypothesis, the evidence from interglacial stage 11 actually supports it.

Summary

This detailed look at the two methods, insolation-alignment and deglacial-alignment, of comparing greenhouse-gas concentrations in past interglaciations against those in the present one supports the early anthropogenic hypothesis. CO_2 and CH_4 values should have been dropping for the last several thousand years. The fact that they have been rising instead is consistent with unprecedented intervention in the climate system, presumably by humans.

Additional Resources

Broecker, W. S., and T. L. Stocker. "The Holocene CO_2 Rise: Anthropogenic or Natural?" *Eos Transactions, American Geophysical Union* 87 (2006): 27.

Crucifix, M., M.-F. Loutre, and A. L. Berger. "Commentary on 'The Anthropogenic Era Began Thousands of Years Ago.'" *Climatic Change* 69 (2005): 419–426.

EPICA Community Members. "Eight Glacial Cycles from an Antarctic Ice Core." *Nature* 429 (2004): 623–628.

Lisiecki, L. E., and M. E. Raymo. "A Plio-Pleistocene Stack of 57 Globally Distributed Benthic $\delta^{18}O$ Records." *Paleoceanography* 20 (2005), PA1003. doi:10.1029/2004PA001071.

Rohling, E. J., K. Braun, K. Grant, M. M. Kucera, A. P. Roberts, M. Siddall, and G. Trommer. "Comparison between Holocene and Marine Isotope Stage-11 Sea-level Histories." *Earth and Planetary Science Letters* 291 (2010): 97–105. doi:10.1016/j.epsl.2009.12.054.

CHAPTER 10

NATURAL VERSUS ANTHROPOGENIC CH₄ SOURCES: CLOSER SCRUTINY

Chapter 9 showed that the rising methane levels during the last 5,000 years differed from the trends in all seven previous interglaciations, when concentrations fell. Although the spread of methane-producing agricultural activities coincides with this wrong-way methane trend, and seems a promising explanation for it, the debate on this issue is not yet entirely settled. This chapter will take a closer look at all possible sources of methane (Tables 10-1 and 10-2).

TABLE 10-1 Changes in Natural CH₄ Sources During the Last 7,000 Years

Source	Change
Arctic wetlands	decreasing
North tropical/Subtropical wetlands	decreasing
South tropical/Subtropical wetlands	increasing

TABLE 10-2 Changes in Anthropogenic CH$_4$ Sources During the Last 7,000 Years

Source	Change
Domesticated livestock	increasing
Rice irrigation	increasing
Biomass burning	increasing
Human waste	increasing

Natural Methane Sources

Recall from Chapter 2 that natural wetlands are the major source of methane emissions and of long-term changes in atmospheric concentrations of CH$_4$ (Figure 10-1). Evidence shows that two major natural wetland regions on Earth could not have been responsible for the CH$_4$ increase during the last 5,000 years:

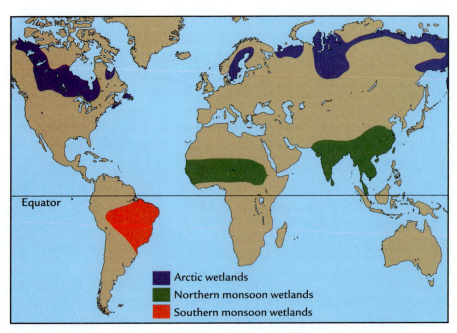

Equator

■ Arctic wetlands
■ Northern monsoon wetlands
■ Southern monsoon wetlands

FIGURE 10-1 The world's three major methane-producing wetland sources: north-tropical monsoon areas, Arctic wetlands, and south-tropical monsoon areas.

1. Evidence from lake levels, pollen records of past vegetation, and geochemical records from $CaCO_3$ deposits in caves show that north-tropical monsoonal wetlands could not have been the source of the additional methane. These records show a broad drying trend extending across a vast region from North Africa, eastward across southern Arabia, and farther eastward through India and into southern China during the last 5,000 years. This strong drying trend indicates that summer monsoons in the Northern Hemisphere were weakening, which would have reduced methane emissions from natural wetlands in the northern tropics.

2. Arctic wetlands, another important methane producer on a global scale, were not responsible, even though these wetlands slowly expanded during the last 5,000 years. The evidence in favor of this conclusion comes from the decreasing difference between CH_4 concentrations in Greenland and Antarctic ice (recall Chapter 2). The diminishing gap between the slightly higher concentration in Greenland ice and the slightly lower concentration in Antarctic ice during the last 5,000 years indicates that far-northern sources near Greenland were emitting progressively less methane. The trend toward cooler summers at Arctic latitudes more than countered the ongoing expansion of Arctic wetlands and thereby reduced total CH_4 emissions.

A South American Explanation for the CH₄ Increase?

The analysis in Chapter 2 omitted a potentially important wetland methane source in the Southern Hemisphere: the Amazon Basin of South America (see Figure 10-1). Like north-tropical monsoon regions, southern Amazonia also has a summer monsoon but at the exact opposite time of the year. The Sun moves into the Southern Hemisphere during the months of northern winter. At the December solstice, the Sun lies directly overhead at 23.5°S, initiating the southern summer. As a result, summer insolation in the Southern Hemisphere occurs in December, January, and February, rather than in June, July, and August, as it does in the north.

Because of these opposed monthly insolation trends, summer monsoons in the Northern and Southern Hemispheres follow opposite trends that vary over long (orbital) time scales (Figure 10-2). For example, 10,000 years ago, when higher than normal June–August insolation was

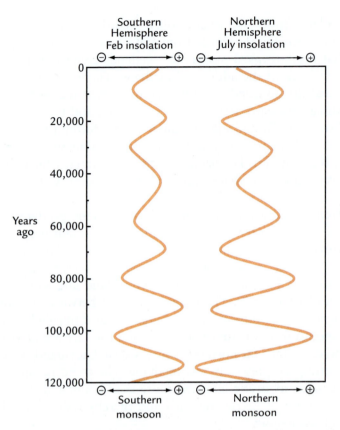

FIGURE 10-2 Orbital insolation trends have opposite monthly timing in the Northern Hemisphere (summer in June, July, and August) and the Southern Hemisphere (summer in December, January, and February). *[Adapted from A. Berger, "Long-term Variations of Caloric Insolation Resulting from the Earth's Orbital Elements,"* Quaternary Research *9 (1978): 139–167.]*

driving unusually strong monsoons in the north, lower than normal December–February insolation was causing unusually weak monsoons in the south (and vice versa).

These opposite responses continued for the last 10,000 years. As June–August insolation weakened in the northern tropics and caused wetlands to dry out, December–February insolation strengthened in the southern tropics. This increase in summer insolation should have driven a strengthening wet summer monsoon that filled wetlands in the Amazon and emitted increasing amounts of methane.

Evidence confirms that the southern monsoon did follow this pattern during the last 10,000 years in Amazonia: lake levels were rising, and wet-adapted vegetation was replacing dry-adapted forms. Oxygen-isotope trends in cave deposits also show increased summer monsoon rains during this interval. In view of this evidence, could expansion of wetlands in southern Amazonia explain the rise in CH$_4$ since 5,000 years ago?

The answer to this question requires determining whether or not the increasing CH$_4$ emissions in the Southern Hemisphere were large enough to offset the decreasing emissions in the north. Several observations suggest that they were not. The southern Amazon Basin is the only large-scale monsoonal wetland region in the Southern Hemisphere, primarily because the rest of the southern tropics lack the large land masses and high mountain topography needed to drive strong summer monsoon circulations. Africa south of the equator tapers in width and is relatively arid, so its summer monsoon is very weak. Australia is a small, mostly arid continent with no large mountain ranges and a weak monsoon only along its northern margin. Even within South America, the (smaller) part of the continent lying north of the equator follows the north-tropical monsoon tempo, with decreasing monsoon strength in the last 10,000 years. As a result, it seems unlikely that total CH$_4$ emissions from the Southern Hemisphere would have been large compared to those in the north.

Several climate models that were used to simulate long-term monsoon responses to changing summer insolation support the claim that southern monsoons have been weaker than those in the north. In one simulation, the area of Amazonia strongly affected by the South American monsoon was smaller than the combined area of the northern monsoon regions by roughly a factor of three. These simulations also indicate that variations in monsoon strength through time were smaller in southern Amazonia than in southern Asia and North Africa by about a factor of two. This combination of smaller monsoon areas and weaker rainfall variations suggests that the relative methane contribution from the southern Amazon would have been smaller than that from the northern monsoons.

These modeling results imply that the weakening of the larger northern monsoons during the last 10,000 years should have outweighed the strengthening of the smaller southern monsoons, causing an overall reduction in the total CH$_4$ concentration in the atmosphere. The known decrease in CH$_4$ emissions from Arctic wetlands would have added to this imbalance.

Additional evidence of a northern control on past methane variations comes from changes during previous interglaciations. CH$_4$ variations during all interglaciations prior to the current one must have been natural in origin, because humans could not have played a significant role in greenhouse-gas changes at those times. The decreases in methane concentrations that occurred in all seven previous interglaciations indicate a northern-dominated response (see Chapter 9, Figure 9-2). In each of those intervals, decreasing summer insolation in the northern tropics weakened the northern monsoons and dried out the regional wetland methane sources, and decreasing summer insolation at higher latitudes caused summer cooling of Arctic wetlands, again suppressing methane emissions. These two processes are thought to explain the falling CH_4 concentrations during the seven previous interglaciations, despite the fact that an opposing increase in monsoon strength was occurring in southern Amazonia. The southern hemisphere contribution at those times was clearly subordinate to that from the north.

On the other hand, a recent study by climate modeler Joi Singarayer and colleagues proposed that natural mechanisms might explain at least a part of the CH_4 increase during the last 5,000 years. Using a model that simulated changes in climate and vegetation types in response to orbital insolation changes, they derived estimates of methane emissions and atmospheric concentrations during the last 130,000 years (Figure 10-3). Their model-simulated CH_4 concentrations broadly matched those recorded in ice cores during most of that interval, consistent with the view summarized above that long-term changes in atmospheric methane are controlled by northern wetlands responding to northern hemisphere insolation changes.

During the transition from the last interglaciation (stage 5) to the subsequent start of the next glaciation, the interval that was most nearly equivalent to the last 5,000 years occurred from 121,000 to 116,000 years ago. For that interval, the model developed by Singarayer and colleagues simulated a decrease in CH_4 concentrations, consistent with the trend in the ice cores (see Figure 10-3). Their model also simulated an increase in methane emissions from the Southern Hemisphere during that time, but a considerably larger decrease in emissions from the Northern Hemisphere. As a result, the model showed that the Northern Hemisphere remained in control of the falling CH_4 concentration in the atmosphere.

But something different happened during the last 5,000 years. Singarayer and colleagues found that their model simulated an increase in methane concentrations, even though northern summer insolation was falling (see Figure 10-3 inset). This decoupling between insolation and simulated methane levels had not occurred during any previous interval in the experiment. Their simulation again produced the expected increase

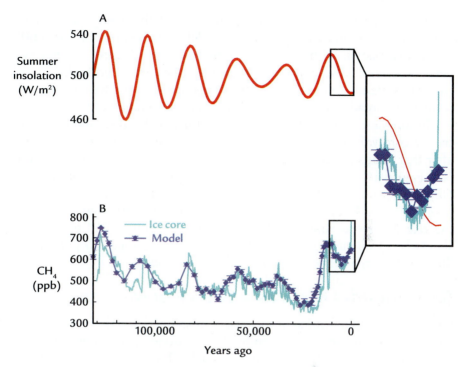

FIGURE 10-3 Observed insolation over the last 130,000 years (A) compared to measured and model-generated estimates of methane concentration (B). Inset area shows trends during the last 10,000 years. *[Adapted from J. Singarayer et al., "Late Holocene Methane Rise Caused by Orbitally Controlled Increase in Tropical Sources," Nature 470 (2011): 82–85, doi:10.1038nature09379.]*

in methane emissions in the Southern Hemisphere and the expected decrease in the north during the last 5,000 years, but the decrease in the north was so small that it failed to outweigh the increase in the south.

Singarayer and colleagues concluded that ice sheets were the reason for these different responses during the two interglaciations. During the latter part of the most recent interglaciation (121,000 to 116,000 years ago), northern ice sheets had begun growing, cooling the Northern Hemisphere enough to drive northern CH₄ emissions to lower levels, and thereby overwhelming the increasing CH₄ emissions from the Amazon.

During the last 5,000 years, however, ice sheets did not grow; only a small cooling occurred, and northern methane emissions in the model barely fell at all. So the natural increase in southern hemisphere emissions took overall control of the global CH₄ level for the first time in the last 130,000 years.

The question remains as to whether or not the negligible decrease in northern CH_4 emissions that their model simulated for the last 5,000 years is consistent with the evidence presented in Chapter 2. That evidence indicated decreasing methane emissions from both the broad tropical wetland region where northern monsoons were weakening, and from widespread Arctic wetlands that were becoming colder.

In any case, the methane increase simulated by their model tracks the ice-core values closely from 5,000 until about 1,000 years ago, but after that it fails to capture the rapid late pre-industrial CH_4 rise observed (see Figure 10-3). The total CH_4 increase they simulated for the last 5,000 years up until the start of the Industrial Era was about 60 ppb (from 585 to 645 ppb), while the increase measured in ice cores through the year 1850 was roughly 200 ppb (from 560 ppb to 760 ppb). Their simulation accounts for about 30% of the total change (60 ppb out of 200), leaving the other 70% of the trend unexplained.

River-Mouth Deltas as a Source of Methane?

One other possible source of increased atmospheric methane during the last few thousand years has been briefly mentioned in the literature—the growth of river-mouth deltas. When sea level stabilized after ice-sheet melting ended 6,000 years ago, deltas began to build out on coastlines in many regions, including China, India, and the Mediterranean coast of Europe and Asia. Many port cities in the Mediterranean had to be relocated seaward several times as their harbors filled with silt (see Chapter 4, Figure 4-14). The expansion of swampy bogs in these deltas emitted methane.

But humans played a significant role in creating these river-mouth deltas (recall Chapter 4). Farming destabilized steep slopes, caused erosion, and sent silts and clays down rivers to delta regions. In the Mediterranean, rates of delta-building increased during times of deforestation. In any case, no one has yet suggested that the expansion of deltas was a major methane source, or at least not one large enough to rival the monsoon regions and Arctic wetlands.

Anthropogenic Methane Sources

Anthropogenic sources are the alternative to natural explanations for the increase in atmospheric methane during the last 5,000 years (see Table 10-2). This possibility (among others) was first noted by the ice-core geochemist Jérôme Chappelaz, but it was not further pursued by him (or others) until the early anthropogenic hypothesis appeared.

Evidence from Eurasia (recall Chapter 5) shows that methane-generating agricultural activities began to increase noticeably around 5,000 years ago, just as the atmospheric CH$_4$ trend reversed direction. Farming then continued to spread across Southeast Asia during subsequent millennia, as the methane concentration rose.

Dorian Fuller and colleagues mapped the spread of rice irrigation across Southeast Asia, India, and North Africa (recall Chapters 5 and 7). By combining the time of initial arrival of rice irrigation in each area with estimates of the subsequent rate of increased density of rice farming within each area as populations grew, they calculated the total increase in irrigated Asian land as a percentage of the modern value (see Chapter 8, Figure 8-11). These estimates of total area irrigated from 5,000 to 1,000 years ago were then converted to estimates of methane emissions, using the modern relationship between total area irrigated and methane emissions as the basis for the calculation.

Because methane emissions control atmospheric CH$_4$ concentrations with minimal lag, Fuller and colleagues converted their estimates of methane emissions from rice irrigation to estimates of methane concentrations in the atmosphere. The trend they calculated showed an exponential rise similar to the measured increase in ice-core CH$_4$ concentrations since 5,000 years ago (Figure 10-4).

According to their calculation, the atmospheric methane contribution from irrigated rice farming had reached 38% of the modern value by the year 1,000. Because rice irrigation today is thought to account for roughly 200 ppb of the total modern atmospheric concentration, their estimate indicates that irrigation had caused a rise of about 76 ppb by the year 1,000 (200 ppb × 0.38 = 76 ppb). This estimate would account for most of the CH$_4$ increase from 5,000 to 1,000 years ago measured in ice cores, roughly 100 ppb (see Figure 10-4).

The analysis to this point has focused on possible causes of the observed CH$_4$ increase, but has ignored the fact that the methane concentration should have fallen during the last 5,000 years as it did during previous interglaciations (see Chapter 9, Figure 9-2). With both factors considered, the full methane anomaly proposed in the early anthropogenic hypothesis is more than twice as large as the observed rise (Figure 10-5). As a result, major additional sources of anthropogenic methane are needed to account for the full proposed anomaly.

One promising source of additional early anthropogenic methane is the spread of livestock that was also mapped by Fuller and colleagues (Chapters 5 and 7). Today, estimated methane emissions from livestock exceed those from rice irrigation, and they may well have

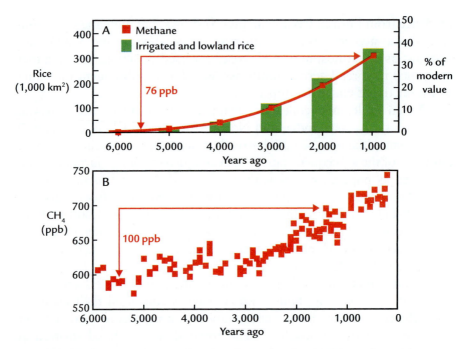

FIGURE 10-4 Estimated area of irrigated rice farming in Asia and resulting contribution to atmospheric CH_4 concentrations from 5,000 to 1,000 years ago (A) compared to observed atmospheric methane levels (B). *[Adapted from D. Q. Fuller, J. Van Etten, K. Manning, C. Castillo, E. Kingwell-Banham, A. Weisskopf, L. Kin, Y.-I. Sato, and R. Hijmans, "The Contribution of Rice Agriculture and Livestock Pastoralism to Prehistoric Methane Levels: An Archaeological Assessment," The Holocene 25 (2011): 743–759, doi:10.1177/0959683611398052.]*

done so in past millennia. Fuller's maps show livestock **pastoralism** beginning to spread across central Asia and northern Africa after 7,000 years ago (Figure 10-6), but the earliest spread occurred mainly in arid and semi-arid regions that have always been sparsely populated and could not have contributed much to global methane emissions. Agriculturalists moving into Europe after about 7,500 years ago also tended livestock (see Chapter 4), but initially their numbers were probably small.

After 5,000 years ago, however, just as the atmospheric CH_4 concentration began to rise, animal husbandry spread rapidly across semi-humid and humid regions of Africa and Asia (see Figure 10-6). By 3,000 years ago, tending livestock occurred in most regions where it does today, except for far southern Africa. As human populations in these

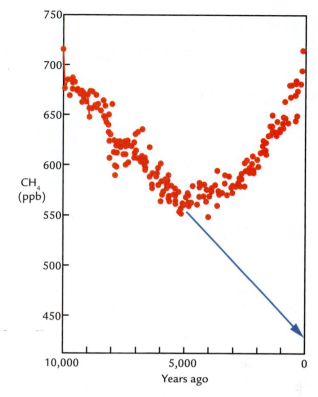

FIGURE 10-5 The full CH₄ anomaly posed by the early anthropogenic hypothesis includes both the observed rise during the last 5,000 years (*red dots*) and the projected fall that did not occur (*blue line*).

regions of ample rainfall began to rapidly increase, livestock populations are likely to have followed a similar trend, with a large resulting effect on methane emissions.

Quantifying the early increases in livestock populations and their effect on methane emissions is difficult because of several complications. As noted in Chapter 8, increased population density and reduced per capita land holdings began to force people to make a nutritional choice between raising livestock and growing crops. Eventually, very high population densities forced people in some areas of China and India to choose intensive crop cultivation at the expense of pasture for livestock. Two other factors, increased biomass burning and greater production of human waste, would have further added to total methane emissions during the last 5,000 years, but, again, the amounts are difficult to estimate.

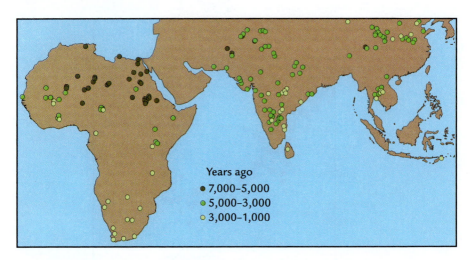

FIGURE 10-6 Sites of first appearance of livestock show the expansion of pastoralism across Asia and Africa. *[Adapted from D. Q. Fuller et al., "The Contribution of Rice Agriculture and Livestock Pastoralism to Prehistoric Methane Levels: An Archaeological Assessment," The Holocene 25 (2011): 743–759, doi:10.1177/0959683611398052.]*

Summary

All four anthropogenic sources of methane listed in Table 10-2 greatly increased during the last 5,000 years. Initial attempts to estimate methane emissions from rice irrigation support an anthropogenic explanation for most of the CH_4 rise between 5,000 and 1,000 years ago, and emissions from the spread of livestock hold the promise of explaining much of the rest of the proposed methane anomaly. In addition, all of the evidence in support of an anthropogenic explanation is consistent with the anomalous rise in methane during the last 5,000 years compared to the downward CH_4 trends in every previous interglaciation (recall Chapter 9).

Further archeological and archeobotanical work, along with greater exploration of historical sources, should help to quantify the link between the growth of anthropogenic sources and the size of the proposed anomaly in methane concentration. If Fuller's pioneering estimate of emissions from rice irrigation is confirmed and supplemented by estimates of the contribution from livestock, natural CH_4 sources may not be necessary to explain the observed CH_4 increase.

Additional Resources

Cruz, F. W., S. J. Burns, I. Karmann, W. D. Sharp, M. Vuille, A. O. Cardoso, J. A. Ferrari, P. L. Silva Dias, and O. Viana, Jr. "Insolation-driven Changes in Atmospheric Circulation over the Past 116,000 Years in Subtropical Brazil." *Nature* 434 (2005): 64–66.

Fuller, D. Q., J. Van Etten, K. Manning, C. Castillo, E. Kingwell-Banham, A. Weisskopf, L. Kin, Y.-I. Sato, and R. Hijmans. "The Contribution of Rice Agriculture and Livestock Pastoralism to Prehistoric Methane Levels: An Archaeological Assessment." *The Holocene* 25 (2011): 743–759. doi:10.1177/0959683611398052.

Kutzbach, J. E., X. Liu, Z. Liu, and G. Chen. "Simulation of the Evolutionary Response of Global Summer Monsoons to Orbital Forcing over the Past 280,000 Years." *Climate Dynamics* 30 (2008): 567–579. doi:10:1007 /s00382-007-0308-z.

Ruddiman, W. F., and J. S. Thomson. "The Case for Human Causes of Increased Atmospheric Methane over the Last 5000 Years." *Quaternary Science Reviews* 20 (2001): 1769–1777.

Ruddiman, W. F., Z. Guo, X. Zhou, H. Wu, and Y. Yu. "Early Rice Farming and Anomalous Methane Trends." *Quaternary Science Reviews* 27 (2008): 1291–1295. doi:10.1016/jquascirev.2008.03.007.

Schmidt, G. A., D. T. Shindell, and S. Harder. "A Note on the Relationship between Ice Core Methane and Insolation." *Geophysical Research Letters* 31 (2004): L23206.

Seltzer, G., D. Rodbell, and S. J. Burns. "Isotopic Evidence for Late Glacial and Holocene Hydrologic Changes in Tropical and South America." *Geology* 28 (2000): 35–38.

Singarayer, J. S., P. J. Valdez, P. Friedlingstein, S. Nelson, and D. J. Beerling. "Late Holocene Methane Rise Caused by Orbitally Controlled Increase in Tropical Sources." *Nature* 470 (2011): 82–85. doi:10.1038/nature09379.

Wang, Y., H. Cheng, R. L. Edwards, Y. He, X. Kong, Z. An, J. Wu, M. J. Kelly, C. A. Dykoski, and X. Li. "The Holocene Asian Monsoon: Links to Solar Changes and North Atlantic Climate." *Science* 308 (2005): 854–857.

Yu, Z., D. W. Beilman, and M. C. Jones. "Sensitivity of Northern Peatland Carbon Dynamics to Holocene Climate Change," in *Carbon Cycling in Northern Peatlands*, edited by A. J. Baird, L. R. Belyea, X. Comax, A. Reeve, and I. Slater, pp. 55–69. Washington, D.C.: American Geophysical Union, 2010. doi:10.1029/2008GM000822.

CHAPTER 11

NATURAL VERSUS ANTHROPOGENIC CO$_2$ SOURCES: CLOSER SCRUTINY

The scientific debate over whether the CO$_2$ trend of the last 7,000 years was natural or anthropogenic continues. Those who favor natural explanations have to account for the 22-ppm CO$_2$ increase since 7,000 years ago, while the anthropogenic explanation must explain a total anomaly of about 40 ppm, the difference between the observed CO$_2$ increase and the decreases typical of previous interglaciations.

Proposed Natural Sources for the 22-ppm CO$_2$ Increase

The two natural explanations that have gained the widest attention both center on changes in the carbon chemistry of the oceans (Table 11-1). Most carbon in the oceans occurs in dissolved forms, either as carbon dioxide (CO$_2$), carbonate ions (CO$_3^{-2}$), or bicarbonate ions (HCO$_3^-$). Over intervals of thousands of years, oceans maintain an approximate balance between dissolved carbon dioxide (which is acidic) and carbonate ions (which are alkaline). When external factors disturb this long-term balance by altering the amount of either carbon dioxide or carbonate ions, the ocean gradually works to restore this equilibrium. Both the oceans and the atmosphere participate in these changes because they exchange large amounts of carbon every year (see Chapter 3).

TABLE 11-1 **Changes in Natural Sources of CO_2 During the Last 7,000 Years**

Source	Change
Ocean carbonate compensation	increasing?
Coral reef building	increasing?

Carbonate Compensation Hypothesis

One natural explanation for the CO_2 increase, proposed by the geochemist Wallace Broecker and colleagues, involves a delayed response in the chemistry of the oceans to changes that had occurred in the final stages of the preceding deglaciation. From 11,000 to 7,000 years ago, forests were growing at high northern latitudes in places where retreating ice sheets had melted. The carbon stored in the wood of these new forests was taken from the combined ocean-atmosphere CO_2 reservoirs (Figure 11-1A). The removal of CO_2 from the ocean between 11,000 and 7,000 years ago left seawater enriched in carbonate ions, which made seawater less acidic, allowing more soft $CaCO_3$ sediment to be deposited on the seafloor. The removal of carbon from the atmosphere caused CO_2 values to fall from 11,000 to 7,000 years ago.

By 7,000 years ago, forests had reached their northern limits and were no longer extracting carbon from the ocean and atmosphere. With more CO_2 left in the ocean, seawater became more acidic and began to dissolve the $CaCO_3$ that had previously been deposited on the seafloor (Figure 11-1B). And more CO_2 was left in the atmosphere, so CO_2 concentrations began to rise. According to rough estimates, this mechanism might have accounted for as much as half of the 22-ppm rise in atmospheric CO_2 since 7,000 years ago. The gradual rise of atmospheric CO_2 concentrations during the last 7,000 years was determined by the slow pace of $CaCO_3$ dissolution on the seafloor. Because the oceans contain much more carbon than the atmosphere does, ocean reservoirs dominate the joint response.

Quantifying the amount of $CaCO_3$ dissolution on the seafloor during the last 7,000 years would seem to provide a test of this hypothesis, but no widely accepted method exists for estimating the amount of seafloor $CaCO_3$ dissolved in the past. Most techniques indicate only the relative sense of change (that is, whether dissolution was increasing or decreasing). In addition, only a handful of carbonate-rich seafloor areas with well-dated cores and high sedimentation rates have been

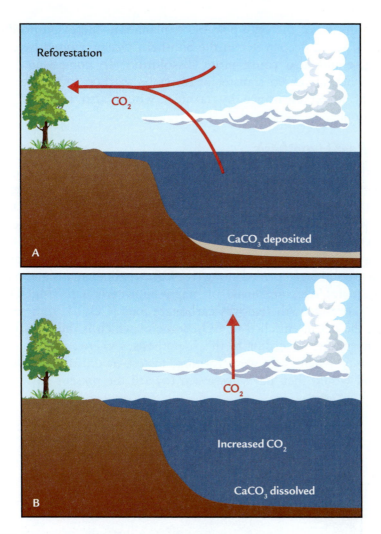

FIGURE 11-1 The **carbonate compensation hypothesis** proposes that forest growth took CO$_2$ carbon from the ocean and atmosphere prior to 7,000 years ago (A), but when forest regrowth stopped, CO$_2$ concentrations in the ocean and atmosphere increased (B).

examined to date, far short of the coverage that would be needed to derive a reliable global estimate. Given the few areas of seafloor suitable for such studies, a full test of the carbonate compensation hypothesis in the future seems unlikely.

Most of the cores that have been examined do show increasing dissolution during the last 7,000 years, but the carbonate compensation mechanism is not the only possible explanation. Increased production of highly acidic deep water in the Southern Ocean, caused by climatic cooling, could also cause more dissolution. To complicate matters even further, increased transfer of terrestrial carbon to the ocean as a result of forest clearance by farmers would also increase $CaCO_3$ dissolution.

Coral Reef Hypothesis

The paleoclimatologist Andrew Ridgwell and colleagues proposed an alternative natural explanation for the CO_2 increase, involving increased construction of coral reefs in tropical regions during the last 7,000 years (Figure 11-2). In most tropical areas, sea level reached its present position (or very near it) when the northern ice sheets finished melting between 7,000 and 6,000 years ago. As sea level stabilized, extensive coral reefs formed along tropical coastlines and island margins. Because coral reefs are made of magnesium carbonate ($MgCO_3$) and $CaCO_3$, coral growth removes carbonate ions (CO_3^{-2}) from seawater, leaving the ocean enriched in CO_2, some of which is then emitted into the atmosphere. According to this hypothesis, CO_2 releases would have been larger when reefs were forming fastest (soon after 7,000 years ago) and would then have decreased toward the present day.

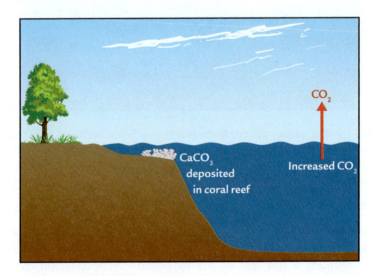

FIGURE 11-2 The **coral reef hypothesis** proposes that stabilization of sea level near its present position 6,000 years ago led to construction of $CaCO_3$ reefs and the release of CO_2 into the surface ocean and atmosphere.

Estimates of the amount of CO_2 that might have been emitted into the atmosphere from the growth of coral reefs have varied widely from 5 to 40 ppm. The actual amount is not known, because only a few tropical reef areas have been intensively studied, and dating enough reefs to characterize the full global response would be an overwhelming task. As a result, scientists do not know whether the growth of tropical reefs actually increased after 7,000 years ago and then slowed toward the present. As was the case for the carbonate compensation hypothesis, the lack of field evidence hinders attempts to evaluate the coral reef explanation for the 22-ppm pre-industrial rise in CO_2 concentration during the last 7,000 years.

Proposed Anthropogenic Sources for the 40-ppm CO_2 Anomaly

The alternative to natural explanations of the CO_2 rise during the last 7,000 years is increased emissions from anthropogenic sources (Table 11-2). In 2002, the ecologist Christopher Carcaillet and colleagues inferred that an increase in the burning of forests indicated by greater concentrations of charcoal fragments in sediments on the continents should have contributed to the rising CO_2 trend, but they did not pursue this idea further.

In 2003, as part of the original early anthropogenic hypothesis, I inferred a 40-ppm CO_2 anomaly and proposed that pre-industrial deforestation was its major source. This interpretation was largely based on a somewhat crude calculation of nearly total deforestation in several heavily-populated regions.

Scientists soon challenged this claim based on a variety of arguments, including the fact that land-use modeling simulations performed up to that time indicated little deforestation prior to the 1800s. These simulations, based on constant per capita land use, suggested that pre-industrial carbon emissions from deforestation could only account for

TABLE 11-2 Changes in Anthropogenic CO_2 Sources During the Last 7,000 Years

Source	Change
Deforestation	increasing (23–24 ppm?)
Coal and peat burning	increasing (2 ppm?)
Ocean CO_2 feedback	increasing (8 ppm or more)

a CO_2 rise of 3 to 5 ppm, just a small part of the observed 22-ppm decrease, and much less than the 40-ppm amount proposed in the early anthropogenic hypothesis.

As noted in Chapter 8, however, those simulations that assumed small and unchanging per capita land use in the past have been shown to be incorrect. That assumption was refuted by Jed Kaplan's studies that used historical data to show that per capita land use (and early deforestation) was much larger in the past. Kaplan's 2009 simulation showed extensive early clearance in Europe, and his 2010 study suggested that the same was true for other densely populated areas: China, India, Mexico, Peru, and parts of Africa.

Kaplan and colleagues estimated 340 billion tons of cumulative carbon emissions just prior to the start of the Industrial Era (Figure 11-3), an amount far higher than the 50–70 billion tons estimated from previous land-use simulations. The release of 340 billion tons of carbon is equivalent to a net increase of atmospheric CO_2 of 24 ppm, again far higher than the 3 to 5 ppm suggested by previous land-use simulations, although still well short of the 40-ppm increase proposed in the original early anthropogenic hypothesis.

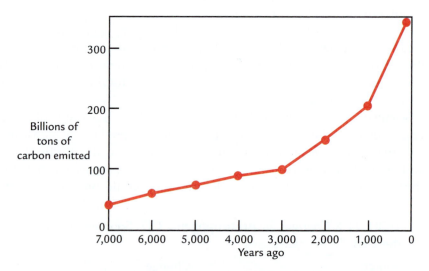

FIGURE 11-3 Estimated cumulative release of carbon from land clearance from 7,000 years ago until 1850. [*Adapted from J. O. Kaplan, K. M. Krumhardt, E. Ellis, W. F. Ruddiman, C. Lemmen, and K. K. Goldewijk, "Holocene Carbon Emissions as a Result of Anthropogenic Land Cover Change," The Holocene 25 (2010): 775–791, doi:10.1177 /0959683610386983.*]

Carbon-Isotope Evidence for Anthropogenic CO$_2$ Emissions

A second potential measure of the amount of early anthropogenic carbon emissions is based on **carbon isotopes** measured in CO$_2$ trapped in ice-core air bubbles. The carbon in atmospheric CO$_2$ has two major sources: (1) inorganic carbon dissolved in the ocean, which is relatively rich in the ^{13}C isotope, and (2) organic carbon from sources on land (mostly forest vegetation), with relatively more of the ^{12}C isotope. Measurements of the ratio of these two kinds of carbon in CO$_2$ from ice-core air bubbles can potentially provide information about how much forest (^{12}C) carbon entered the atmosphere over specific intervals in the past.

Analyses of carbon-isotope changes by scientists at the University of Bern indicate a total release of about 50 billion tons of terrestrial carbon over the last 7,000 years. This number represents the net sum of several processes, some of which cause carbon emissions from sources on land, and others that cause carbon to be stored in natural terrestrial "reservoirs." As part of their analysis, the Bern group estimated that roughly 30 billion tons of ^{12}C-rich carbon were released from the land to the atmosphere by the weakening of northern monsoons and the resulting loss of carbon-rich vegetation during the last 7,000 years, while another 50 billion tons were emitted from deforestation by humans (Figure 11-4A).

Their analysis also assumed that these emissions were partially offset by the storage of some 40 billion tons of carbon buried in Arctic peat bogs during the last 7,000 years. Removing ^{12}C-rich carbon from the atmosphere and storing it in peat bogs reduces the net emission of forest carbon recorded by the carbon-isotope analyses. The Bern group estimate of 50 billion tons of anthropogenic deforestation is equivalent to a 3.5-ppm increase in atmospheric CO$_2$, which matches initial estimates from early land-use simulations, and falls far short of the 40 ppm proposed in the anthropogenic hypothesis.

But this analysis is vulnerable to a serious criticism: the Bern group overlooked credible published estimates suggesting a much larger amount of carbon burial in peat bogs during the last 7,000 years—roughly 300 billion tons. If this larger estimate is correct, it would lead to a very different interpretation of the carbon-isotope record (see Figure 11-4B). In this new interpretation, the overall carbon-isotope evidence for a *net* release of about 50 billion tons from terrestrial carbon still holds, but now the much larger removal of carbon buried in peat bogs requires a comparably larger offsetting source of terrestrial emissions. Because the estimate of carbon emissions from natural processes (such as monsoons) remains nearly the same, the only viable source is emissions of carbon

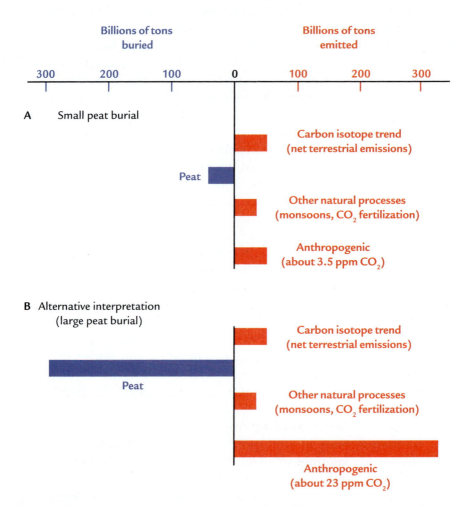

FIGURE 11-4 Two estimates of carbon transfers among major reservoirs that account for a net release of about 50 billion tons of carbon to the atmosphere. *[(A) from J. Elsig et al., "Stable Isotope Constraints on Holocene Carbon Cycle Changes from an Antarctic Ice Core," Nature 461 (2009): 507–510, doi:10.1038/nature.08393; (B) from W. F. Ruddiman, J. E. Kutzbach, and S. J. Vavrus, "Can Natural or Anthropogenic Explanations of Late-Holocene CO_2 and CH_4 Increases Be Falsified?" The Holocene 25 (2011): 793–801, doi:10.1177/0959683610387172.]*

caused by humans deforesting the land. In this alternative interpretation, anthropogenic emissions during the last 7,000 years would have been slightly greater than 300 billion tons of carbon, equivalent to an increase in atmospheric CO$_2$ of about 23 ppm. This estimate falls very close to the value from Kaplan's land-use simulation (about 24 ppm).

Other smaller sources of CO$_2$ emissions may have contributed to the anthropogenic total (see Table 11-2). By 2,000 years ago, people in China had begun to burn coal, which was readily available in small open-pit surface mines. Coal was used by tens of millions of people for cooking and heating houses during the cold winters in northern China (see Chapter 5). Also, during the last few centuries, peat was widely burned in northern Europe for cooking and heating, because wood was no longer easily available from the deforested landscape.

Crude "back-of-the-envelope" estimates (multiplying the number of families by the amount of coal or peat used every year) suggest that each of these factors had only a minor effect on atmospheric CO$_2$ concentrations, probably 1 ppm or less, although these estimates need more careful assessment. In any case, deforestation remains by far the major anthropogenic factor that caused CO$_2$ concentrations to rise. Taken together, the direct effects of human actions might account for about 25 ppm of the proposed 40-ppm CO$_2$ anomaly (see Table 11-2), leaving roughly 15 ppm unexplained.

Feedbacks to Anthropogenic CO$_2$ Emissions

In 2007, I suggested another likely anthropogenic CO$_2$ contribution that had not been part of the original hypothesis: CO$_2$ feedback from the ocean (Figure 11-5). Put simply, both the 25 ppm from direct anthropogenic CO$_2$ emissions and the 250 ppb or more from anthropogenic methane emissions (recall Chapter 10) kept the atmosphere warmer than it would otherwise have been, and the warmer atmosphere warmed the oceans. Warmer oceans would then have released CO$_2$ into the atmosphere, as an indirect anthropogenic feedback.

Two main sources of feedback from the ocean are plausible. The first feedback is the well-understood effect of temperature on the solubility of CO$_2$ in seawater (Figure 11-6). As ocean water warms, it emits more and holds less CO$_2$, similar to how a warm can of soda emits more CO$_2$ bubbles than a cold can does. Two independent methods can be used to estimate the size of this ocean warming caused by changes in **CO$_2$ solubility**. One method uses climate model simulations that allow the ocean to react to the overlying atmosphere to show how much warmer the ocean water (both surface and deep) would have become

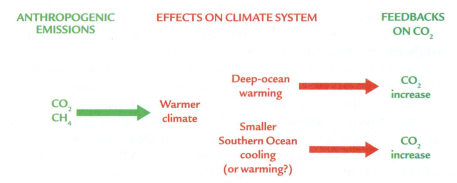

ANTHROPOGENIC
EMISSIONS

EFFECTS ON CLIMATE SYSTEM

FEEDBACKS
ON CO_2

FIGURE 11-5 Proposed CO_2 ocean feedbacks: warming of the atmosphere by anthropogenic CO_2 and CH_4 emissions also warms the oceans, which returns CO_2 to the atmosphere.

because of the extra greenhouse gases produced by early farmers. The second method uses the geochemical (oxygen-isotope) index preserved in the $CaCO_3$ shells of ocean-dwelling organisms (see Chapter 1). During times when ice volume is not changing significantly, as has been the case for the last 7,000 years, changes in this index record changes in deep-ocean temperature.

Both methods converge on a similar answer: human-generated greenhouse gases would have warmed the ocean by about 0.8°C.

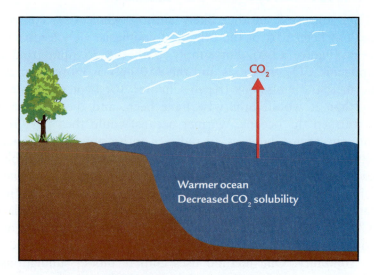

FIGURE 11-6 A warmer ocean holds less dissolved CO_2 than a cool one, and emits more of the greenhouse gas to the atmosphere.

An ocean warmed by this amount would have expelled enough CO$_2$ into the atmosphere to drive its CO$_2$ concentration up by about 8 ppm. This feedback, a direct consequence of human action, would close about half of the remaining gap needed to explain the 40-ppm estimate for the CO$_2$ anomaly (25 + 8 = 33 ppm; see Table 11-2).

Another potential feedback that has not yet been explored involves the Southern Ocean, widely recognized as an important player in the CO$_2$/climate system. When the Southern Ocean cools, more sea ice forms and the overlying wind patterns shift to the north. These changes reduce CO$_2$ exchanges with the overlying atmosphere, lowering its CO$_2$ concentration.

A warmer ocean does just the opposite, leading to greater CO$_2$ exchanges with the atmosphere and higher atmospheric concentrations (Figure 11-7). The size of this effect has not yet been quantified, but an anthropogenically warmed Southern Ocean would likely have contributed CO$_2$ back into the atmosphere, closing part of the remaining gap with the original 40-ppm anomaly. Climate models are just now reaching the point of attempting to simulate the size of all possible CO$_2$ feedbacks from the ocean.

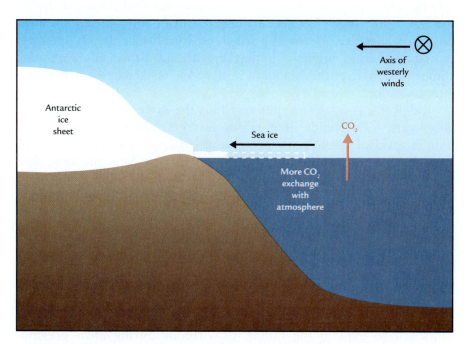

FIGURE 11-7 A warmer Southern Ocean with less sea ice exchanges more CO$_2$ with the overlying atmosphere and causes higher CO$_2$ concentrations.

Previous Interglaciations: Tests of Natural and Anthropogenic Explanations

Both natural sources (see Table 11-1) and anthropogenic sources (see Table 11-2) remain viable explanations of the CO_2 increase during the last 7,000 years. Another way to evaluate these two contending explanations is to compare the rising CO_2 trend in recent millennia with the CO_2 responses during previous interglaciations (see Chapter 9). This comparison provides strong evidence against the two natural explanations, because CO_2 trends during most previous interglaciations headed downward, not upward, contrary to what natural explanations predict.

The natural driving factors Broecker proposed in his carbonate compensation hypothesis were essentially the same during the earlier interglaciations as they were during the current one. In six of the seven previous interglaciations (except for stage 15), ice sheets had melted and forests shifted north, taking carbon out of the atmosphere and ocean. According to the carbonate compensation hypothesis, this sequence of events should have caused a subsequent rise in atmospheric CO_2 concentrations similar to the one that occurred during the last 7,000 years. Yet six of the seven prior interglaciations show no sign whatsoever of an upward trend (recall Chapter 9), contrary to the prediction implicit in that hypothesis. Broecker's carbonate compensation hypothesis fails six of seven tests based on previous interglaciations.

Ridgwell's coral reef hypothesis fails the same test because of the downward CO_2 trends in previous interglaciations. Because of its long interval of stable peak-interglacial sea level (see Chapter 9), the stage 11 interglaciation is particularly instructive. After the long, slow preceding deglaciation ended, sea level stayed high during the peak stage 11 interglaciation between approximately 405,000 and 400,000 years ago, the time equivalent to the stable sea levels during the current interglaciation (see Chapter 9, Figure 9-7). According to the coral reef hypothesis, this time of stable sea level during stage 11 should have led to reef construction that caused a CO_2 increase, but CO_2 did not rise (see Chapter 9, Figure 9-3), so the coral reef hypothesis fails this test. Both the carbonate compensation and the coral reef explanations for the CO_2 increase during the last 7,000 years fail to explain the falling CO_2 trends during most previous interglaciations.

In contrast, the downward CO$_2$ trends during most previous inter-glaciations are in full accord with the anthropogenic explanation, as are the pioneering simulations by Jed Kaplan and colleagues suggesting much larger pre-industrial clearance and carbon emissions than in previous land-use simulations. Larger emissions from land use in turn require larger feedback from the ocean. Future land-use simulations will likely be improved by using new archeological and historical data as additional evidence of land use, deforestation, and CO$_2$ releases.

Remaining Mismatch between Carbon Emissions and CO$_2$ Concentrations

Although the effects of larger early per capita land use boosted early carbon emissions relative to the late-rising population trend (Figure 11-8), the carbon emissions trend still has a late-rising shape that lags well behind the very early rise in CO$_2$ (Figure 11-9). Two additional factors may help to reconcile this mismatch.

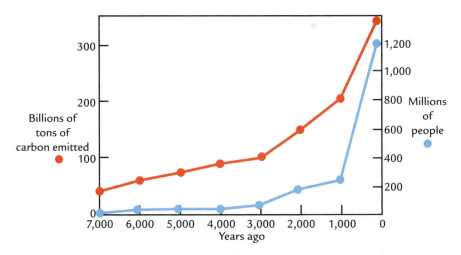

FIGURE 11-8 The estimated cumulative release of carbon from land clearance from 7,000 years ago until 1850 occurs earlier than an estimate of population levels based on the assumption of geometric growth. *[Carbon emissions adapted from J. O. Kaplan, K. M. Krumhardt, E. Ellis, W. F. Ruddiman, C. Lemmen, and K. K. Goldewijk, "Holocene Carbon Emissions as a Result of Anthropogenic Land Cover Change,"* The Holocene 25 *(2010): 775–791, doi:10.1177/0959683610386983. Population growth adapted from C. McEvedy and R. Jones,* Atlas of World Population History *(New York: Penguin Books, 1978).]*

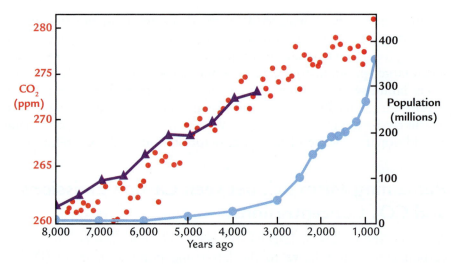

FIGURE 11-9 Atmospheric CO_2 concentration during the last 8,000 years compared to differing estimates of past population. *Light blue circles* are estimates from geometric growth; *dark blue triangles* are estimates from logistical growth. [*CO_2 concentrations from EPICA Community Members, "Eight Glacial Cycles from an Antarctic Ice Core," Nature 429 (2004): 623–628. Population values from J. F. Boyle, M.-J. Gaillard, J. O. Kaplan, and J. A. Dearing, "Modelling Prehistoric Land Use and Carbon Budgets: A Critical Review," The Holocene 25 (2011): 715–722, doi:10.1177 /0959683610386984.]*

Different Population Trends

Estimates of total land clearance and carbon emissions also depend crucially on prehistoric population levels that are not actually known. As noted in the introduction to Part 2 of this book, estimates of past population back to 2,000 years ago are reasonably reliable for China and Europe, but not well known for other regions. Prior to 2,000 years ago, population estimates are basically just informed guesses based on hypothetical assumptions about humans and their environment. Estimates from different demographic models yield dramatically different estimates of early populations (as well as the amount of land they cleared).

The most frequently cited demographic models start with the imperfect estimates from 2,000 years ago and project back in time by assuming **geometric growth** in global and regional populations (roughly a doubling every 1,000 years). This method implicitly assumes that the factors that controlled population growth remained constant through those many millennia and that populations increased by the same constant fraction of the total number of people living at any one time. The assumed growth pattern is reversed to work backward and calculate past populations.

Population estimates derived using this method produce late-rising exponential trends with relatively small populations millennia ago. The most frequently cited reconstruction using the geometric method projects a global population of only 14 million people 5,000 years ago (see Figure 11-9), at a time when the increase in CO_2 concentrations had already been rising rapidly.

In contrast, a second group of demographic models rejects the idea that the fractional growth rate remained constant for those thousands of years of prehistoric time. These alternative estimates assume **logistical growth,** in which early increases in human populations were rapid because natural resources were so abundant that they did not act as a serious check on human expansion. In addition, the relatively small size of populations scattered across early villages and towns limited the effects of epidemics in culling populations.

In later millennia, however, the growing number of people began to encounter resource limitations that made it more difficult to obtain food and slowed the rate of population growth. In addition, as people began to crowd into urban areas, the growing incidence of disease (epidemics and later pandemics) culled more of the population. As a result of these two factors, the rate of population growth slowed.

In the example of a logistical growth model shown by dark blue triangles in Figure 11-9, the estimated human population 5,000 years ago had already reached nearly 200 million people, or almost 14 times the size of the estimate based on geometric growth. This early-rising population estimate plots almost directly in line with the upward CO_2 trend. If this estimate is correct, the previous mismatch between the CO_2 increase and population growth would be eliminated. Other estimates based on differing assumptions about the controls on population have produced estimates intermediate between the two extremes shown in Figure 11-9.

Indirect support for rapid early increases in population comes from the explosive expansion of archeological sites in China between 8,000–7,000 years ago and 5,000–4,000 years ago (see Chapter 5, Figure 5-5). In addition, a very recent study of European DNA indicates that the fastest rate of growth during prehistoric time (calculated as a fraction of the existing population) occurred between 6,500 and 6,000 years ago, near the time of full adoption of agriculture across that continent. Fractional growth rates slowed in subsequent millennia, until the dramatic reduction of mortality rates in recent centuries due to modern medicine and sanitation.

These two lines of evidence suggest that prehistoric global population growth did not follow the widely cited geometric model, with its late-rising form. Instead, population growth appears to have followed the early-rising logistical growth model, which more closely resembles the early-rising shape of the trend in atmospheric CO_2 concentration.

Fire

Another potential factor in early land clearance and carbon emissions is the use of fire. Scientists generally infer that both pre-agricultural people (hunter-gatherers) and early farmers (especially pastoralists) used fire more extensively than later agriculturalists. Before farming, hunter-gatherer people used seasonal burning to maintain grass clearings in order to attract wild game and allow berries and nuts to grow. Historical anecdotes based on encounters of Europeans with early Americans (see Chapter 6) describe their extensive use of fire, but experts continue to disagree about the extent of burning.

Studies of fires ignited by hunter-gatherer people who have lived well way from major population centers in recent decades also support the idea of larger per capita burning by early farmers. At very low population densities (less than ten people per square kilometer), fires set by humans increase directly as populations grow. Fire frequency (per capita) reaches a maximum at population densities near fifteen people (three families) per square kilometer and then declines, gradually trending toward very few fires at very high population densities. Applied to global population estimates over the last few thousand years, this relationship suggest that the clearance per person by fires would have been much larger thousands of years ago, when population densities were very low, than in recent millennia. (The combination of maliciously set fires and climate change in very recent times may have upset this relationship.)

Reliable evidence from New Zealand provides an unusually well-quantified case example (recall Chapter 7). The Maori people, who first arrived around the year 1280, burned forests at the extraordinary rate of about 90 hectares per person, more than 10 times higher than the per capita land-use estimates in Kaplan's simulations (see Chapter 8, Figures 8-6 and 8-8). Several scientists have also argued that the hunter-gatherer people who first arrived in Australia near 45,000 years ago used fire extensively and inadvertently drove most of the wild game (large marsupials) to extinction. And in cases in which early farmers entered previously uninhabited islands such as Madagascar and small islands in the western and equatorial Pacific, their extensive use of fire caused major deforestation and contributed to widespread extinctions of animals. Scientists have much to learn about the effects of fire on early land clearance, but the potential remains for effects much larger than those now included in land-use models.

Summary

Both the natural and anthropogenic explanations for the CO_2 increase during the last 7,000 years remain viable, but the balance of evidence favors the early anthropogenic hypothesis. Early land-use simulations that indicated small pre-industrial carbon emissions at first favored natural explanations, but more realistic simulations based on historical data have boosted pre-industrial emissions toward the level proposed in the early anthropogenic hypothesis. Similarly, initial interpretations of carbon-isotopic trends in CO_2 gas favored natural explanations, but more realistic estimates of large carbon burial in Arctic peat have boosted emissions toward anthropogenic levels. In addition, CO_2 feedback from anthropogenically warmed oceans further increases the total carbon dioxide concentration to levels closer to the 40-ppm value proposed in the original anthropogenic hypothesis. Finally, the lack of upward CO_2 trends during six of seven previous interglaciations favors the early anthropogenic hypothesis.

A definitive resolution of the natural versus anthropogenic debate over the cause of the CO_2 increase during the last 7,000 years awaits several advances: more detailed historical and archeological evidence to refine models of past land use; further attempts to determine the amount of carbon buried in high-Arctic peat bogs; and more detailed DNA evidence to define the times of fastest population growth rates during the last 7,000 years.

Additional Resources

Broecker, W. S., E. Clark, D. C. McCorckle, T.-H. Peng, I. Hajdas, and G. Bonani. "Evidence for a Reduction in the Carbonate Ion Content of the Deep Sea During the Course of the Holocene." *Paleoceanography* 14 (1999): 744–752. doi:10.1029/1999PA900038.

Broecker, W. S., and T. F. Stocker. "The Holocene CO_2 Rise: Anthropogenic or Natural?" *EOS Transactions, American Geophysical Union* 87 (2006): 27.

Elsig, J., J. Schmitt, D. Leuenberger, R. Schneider, M. Eyer, M. Leuenberger, F. Joos, H. Fischer, and T. F. Stocker. "Stable Isotope Constraints on Holocene Carbon Cycle Changes from an Antarctic Ice Core." *Nature* 461 (2009): 507–510. doi:10.1038/nature.08393.

Gajewski, K., A. Viau, M. Sawada, D. Atkinson, and S. Wilson. "*Sphagnum* Peatland Distribution in North America and Eurasia During the Past 21,000 Years." *Global Biogeochemical Cycles* 15 (2001): 297–310.

Gignoux, C. R., B. M. Henn, and J. L. Mountain. "Rapid, Global Demographic Expansions After the Origins of Agriculture." *Proceedings of the National Academy of Sciences* 108, no. 15 (April 2011): 6044–6049. doi:10.1073/pnas.0914274108.

Gorham, E. "Northern Peatlands: Role in the Carbon Cycle and Probable Responses to Climatic Warming." *Ecological Applications* 1 (1991): 182–195.

Joos, F., S. Gerber, I. C. Prentice, B. L. Otto-Bleisner, and P. J. Valdes. "Transient Simulations of Holocene Atmospheric Carbon Dioxide and Terrestrial Carbon Since the Last Glacial Maximum." *Global Biogeochemical Cycles* 18 (2004): GB2002. doi:10.1029/2003GB002156.

Kaplan, J. O., K. M. Krumhardt, and N. Zimmerman. "The Prehistoric and Preindustrial Deforestation of Europe." *Quaternary Science Reviews* 28 (2009): 3016–3034. doi:10.1016/j.quascirev.2009.09.028.

Kaplan, J. O., K. M. Krumhardt, E. C. Ellis, W. F. Ruddiman, C. Lemmen, and K. K. Goldewijk. "Holocene Carbon Emissions as a Result of Anthropogenic Land Cover Change." *The Holocene* 25 (2010): 775–791. doi:10.1177/0959683610386983.

Kutzbach, J. E., S. J. Vavrus, W. F. Ruddiman, and G. Philippon-Berthier. "Comparisons of Coupled Atmosphere-Ocean Simulations of Greenhouse Gas-Induced Climate Change for Pre-Industrial and Hypothetical 'No-Anthropogenic' Radiative Forcing." *The Holocene,* 25 (2011): 793–801. doi:10.1177/095968361038983.

Ridgwell, A. J., A. J. Watson, M. A. Maslin, and J. O. Kaplan. "Implications of Coral Reef Buildup for the Controls on Atmospheric CO_2 Since the Last Glacial Maximum." *Paleoceanography* 18 (2003): 1083. doi:10.1029/2003PA000893.

Ruddiman, W. F., J. E. Kutzbach, and S. J. Vavrus, "Can Natural or Anthropogenic Explanations of Late-Holocene CO_2 and CH_4 Increases Be Falsified?" *The Holocene* 25 (2011): 793–801. doi:10.1177/0959683610387172.

Stephens, B. B., and R. F. Keeling. "The Influence of Antarctic Sea Ice on Glacial/Interglacial CO_2 Variations." *Nature* 404 (2000): 171–174.

Toggweiler, J. R., J. L. Russell, and S. R. Carlson. "Midlatitude Westerlies, Atmospheric CO_2, and Climate Change During the Ice Ages." *Paleoceanography* 21 (2006). doi:10.1029/2005PA001154.

Wirtz, K. W., and C. Lemmen. "A Global Dynamic Model for the Neolithic Transition." *Climatic Change* 59 (2003): 333–367.

Yu, Z. "Holocene Carbon Flux Histories of the World's Peatlands: Global Carbon-Cycle Implications." *The Holocene* 25 (2011): 761–774. doi:10.1177/0959683610386982.

Part 3 Summary

Several kinds of evidence have been brought to bear on the debate over whether the CO_2 and CH_4 increases of the last several thousand years were natural or anthropogenic in origin (Table 3s-1).

TABLE 3s-1 Evidence Favoring Natural and Anthropogenic Explanations for the Greenhouse-Gas Rises in the Last Several Thousand Years

Type of Evidence	Explanation	Favored
	CO_2	CH_4
Historical	Anthropogenic	Anthropogenic
Archeological	[Anthropogenic?]	Anthropogenic
Model simulations	Natural > Anthro*	Mixed
Previous interglaciations	Anthropogenic	Anthropogenic

*Initial simulations favored a natural explanation; later, better-justified ones favor the anthropogenic explanation.

Historical data suggest extensive early land clearance consistent with the early anthropogenic explanation (see Chapters 8, 10, and 11). Although systematic land-use data sets at national levels are rare prior to the 1960s, sparse historical data from China and Europe spanning the last 2,000 years show that past per capita clearance occurred at rates considerably higher than values in recent centuries. These higher early clearance rates indicate large early CO_2 emissions from anthropogenic sources. Similarly, higher early per capita land use in rice-farming regions favors the anthropogenic explanation for the CH_4 rise.

Archeological data indicate the early expansion of methane-producing activities (see Chapters 8 and 10). Methane emissions from the expansion of irrigated rice paddies appear to account for about 75% of the CH_4 increase between 5,000 and 1,000 years ago. Additional CH_4 contributions from the spread of livestock and from other farming activities will add to the anthropogenic total, once they are better quantified by archeologists. At this point, archeological data do not permit the quantification of early CO_2 emissions, but researchers have generally inferred extensive clearance around major population centers by 2,000 years ago (see Chapter 8, Figure 8-8), which is consistent with the anthropogenic hypothesis but not with the more pristine world implied by natural explanations.

Models that simulate climate and vegetation responses to human activities have yielded conflicting results. Models that assumed small constant per capita clearance yielded late-rising carbon emissions trends that agreed with natural explanations for small pre-industrial CO_2 increases. But models based on the large early per capita clearance evident in historical data suggest earlier-rising carbon emissions that are more in agreement with the early anthropogenic hypothesis (see Chapters 8 and 11). Models generally indicate a northern hemisphere control of natural variations in atmospheric methane and favor an anthropogenic origin for the pre-industrial CH_4 increase, but one model favors a (natural) southern hemisphere origin for this CH_4 rise (see Chapter 10).

Comparisons of greenhouse-gas trends during the last several thousand years against trends in previous interglacial intervals favor the anthropogenic explanation. CH_4 trends fell in all seven previous interglaciations, and CO_2 concentrations decreased in six of seven cases, but both trends rose during this current interglaciation. Because previous interglaciations occurred when humans did not play an important climate role, these earlier downward trends can be assumed to represent the natural behavior of the climate system. Compared to these natural trends, the CO_2 and CH_4 increases in recent millennia are anomalous and thus likely anthropogenic.

How Science Moves Forward

No one-size-fits-all description can cover the ways in which scientific advances occur. Some advances come about from meticulous lab work. Others arise from the innovative development of new scientific methods. Some occur through unexpected flashes of insight that detect previously unseen patterns in old data. And, at times, advances occur through serendipity, when a scientist exploring one issue unexpectedly stumbles onto another.

In a general sense, science moves forward by the back-and-forth interaction between observations and ideas. New observations entail steady, patient efforts that are often dismissed as "incremental science," because they nudge knowledge ahead in small steps that are rarely dramatic. Even so, incremental science underpins much of the forward motion of science by continuously adding to the sum total of known "facts." In addition, old data are often reexamined and found to be useful in unexpected ways. Without new facts, science would stall in endless unresolved arguments.

Innovations of new techniques are less common, but especially important. Parts 1–3 of this book used information from a wide range of remarkable new scientific techniques developed and applied in the last few decades: drilling long sequences of cores several kilometers deep into ice sheets; analyzing minute amounts of greenhouse gases (in parts per million or even billion!) from tiny bubbles of ancient air trapped in the ice; retrieving and dating long sequences of marine sediment cores from ocean depths of 3 to 5 kilometers; and refining computer programs that allow astronomers to calculate subtle variations in Earth's orbit hundreds of thousands of years ago.

Many new methods have emerged from the field of geochemistry, including techniques to measure different kinds (isotopes) of oxygen and carbon in marine shells, terrestrial vegetation, and ice-core gases, as well as improved radiocarbon (^{14}C) dating methods that provide time scales for records from the last 10,000 years and earlier. These new methods have driven scientific advances by revealing when events and processes occurred, such as the rates of movement of important substances among the different parts of the climate system.

New hypotheses also play a role in this process. Even though new ideas may be thought to be "a dime a dozen," ideas judged to be bogus either never gain recognition in the larger scientific community or quickly fade away into obscurity. A much smaller subset of ideas (hypotheses) with greater credibility may survive as worthwhile challenges to our ever-changing understanding of science, and an occasional one may earn the higher status of becoming a **theory**.

Historians interested in the workings of the scientific process have looked closely at the great revolutions in knowledge during the last 500 years, such as Copernicus' challenge to the idea of an Earth-centered universe, or geophysical measurements during the mid-1900s indicating that continents must have moved across Earth's surface. Because of resistance from organized religion of medieval and later times, the Copernican revolution took centuries to be fully accepted by the main-stream scientific community. In contrast, after the idea of continental drift had been rejected for half a century, an explosion of observational evidence supporting the newer plate-tectonic explanation swept away the opposition within just a few years during the late 1960s and early 1970s.

In the end, science does move forward, but sometimes only after many twists and turns. Unexpected new evidence can emerge at any time, and older evidence may come to be understood in a different way. Because science relies on the rational interpretation of observations, the eventual outcome of any scientific debate is hard to predict until enough facts have been assembled and carefully weighed by the community.

Part 3 of this book summarized a decade of efforts to evaluate the natural and anthropogenic explanations for the CO_2 and CH_4 increases during the last several thousand years. Given the wide range of evidence covered in Parts 1–3, is it possible to project how the larger scientific community will decide which hypothesis is valid?

Because I am a proponent of one of the competing hypotheses, my assessment in this section should be viewed with healthy skepticism. Nevertheless, my conclusions in this section can be weighed against the wide range of evidence summarized in Parts 1–3. What follows in the next three chapters is the most objective assessment I could make with the evidence at hand.

One working model of how science advances is by **falsification**, in which scientific evidence that emerges after publication of a hypothesis contradicts one or more of its important predictions. Chapter 12 summarizes the evidence reviewed in Part 3 and points to evidence that may be viewed as strong enough to falsify one of the two explanations for the greenhouse-gas increases of the last few thousand years.

Another conceptual model of the scientific process is the concept of a **paradigm shift**: a relatively abrupt transition that occurs when the scientific community comes to the conclusion that a widely accepted explanation (a paradigm) for some observed phenomena is critically flawed and the community abandons that explanation in favor of a new one. Chapter 13 reviews recent assessments of past human effects on climate and weighs the possibility that a paradigm shift favoring large early human effects on the land, the environment, and the climate may be about to occur.

If the early anthropogenic hypothesis were to cause a paradigm shift in how scientists view the long-term history of human influences on climate, what would the new paradigm look like? Chapter 14 summarizes the three major potential changes: (1) the interval of greatest global forest clearance would be pushed back from the Industrial Era (the last 150 years) to prior millennia; (2) the start of large human effects on CO_2 and CH_4 concentrations would similarly be shifted back thousands of years; and (3) the pre-industrial effect of anthropogenic greenhouse gases in warming global climate would be seen as comparable to, or even larger than, the anthropogenic warming of the last 150 years. The current paradigm that Earth's climate stayed warm for natural reasons as civilizations emerged during the last several thousand years would be replaced by a new paradigm in which pre-industrial greenhouse-gas emissions from the spread of early agriculture kept the planet considerably warmer than it would have been had Nature prevailed.

CHAPTER 12

FALSIFICATION

As Stephen Schneider pointed out in the title of his last book, *Science as a Contact Sport,* climate science is not a process gracefully pursued behind ivy-covered academic walls, but a real-world, out-in-the-open battle over ideas. This confrontational approach becomes especially intense when new ideas depart substantially from a conventional view that has long been widely accepted. When new ideas arise, some scientists see them as interesting enough to be worth testing. This inevitable scrutiny is the way science works; no new idea of any importance stands unchallenged for long. In the long run, this built-in skepticism on the part of scientists is a major reason for the enormous advances of the last few centuries, advances that have produced most of the conveniences and comforts that readers of this book currently enjoy (such as electricity, cars, and the Internet).

"Proving" a new hypothesis is not really possible. The chance always exists that later observations may show that the idea doesn't work and should be rejected. Instead, as the Austrian/British philosopher Karl Popper argued, the process of choosing among competing ideas is one of falsification: finding evidence incompatible with predictions implicit in a proposed hypothesis. Many scientists (including me) see falsification as an essential part of the scientific process. In the end, very few hypotheses survive to become recognized as theories.

But is the falsification process absolute and foolproof? Some scientists use Popper's view to claim that even a single line of evidence can falsify an existing hypothesis. And at times a particular falsification can indeed seem so overwhelming that it can convince scientists not to give the new hypothesis under question further consideration. In actuality, however, the falsification process is not absolute, because it depends on judgments made by scientists who may unknowingly lack enough facts to make the claims they are advancing. Evidence based on techniques or measurements that might at first have seemed sufficient to falsify a particular hypothesis may later turn out to fall short, because the conclusions that were drawn depended on invalid assumptions. In such cases, the supposedly falsified hypothesis may be brought back to life upon later reexamination.

In the end, the process of evaluating hypotheses requires scientists to examine all of the evidence, consider all of the competing explanations, and come to their best judgments about the preponderance of information available at the time. But this description of the ideal scientific process omits an important real-world consideration. It implicitly assumes that scientists have enough time to closely monitor issues that lie within or near their own specialties to come to a well-informed judgment. In the real world, however, with so many demands placed on active scientists, this ideal is rarely achieved. Instead, the views of a relatively small subset of scientists tend to dominate the discussion and debate over new hypotheses. This is especially true for strongly interdisciplinary debates such as the one over the early anthropogenic hypothesis.

The central question of this chapter is whether or not the evidence available at the present time is strong enough to falsify either the natural or the anthropogenic explanations for the CO_2 and CH_4 rises of the last several thousand years. Because I proposed the anthropogenic hypothesis, my impartiality and judgment on this issue are obviously suspect. Yet the wide range of evidence from studies reviewed in Part 3 points to several conclusions that stand on their own, reasonably independent of my own opinion.

Falsifying Explanations for the Rising Methane Trend

Nature's own experiments falsify natural explanations for the CH_4 increase of the last 5,000 years (see Chapters 2, 9, and 10). The lack of even a single upward methane trend among the past seven interglaciations (see Chapter 9, Figure 9-2) is a strong falsification of any natural explanation for the recent upward trend.

Because the shapes of the downward CH_4 trends vary widely among the prior interglaciations, the methane trend during the last 10,000 years can be compared to a stacked average of the previous trends (Figure 12-1). For the early part of the current interglaciation, methane concentrations generally followed the averaged trend from previous interglaciations, staying within one standard deviation of the mean value. But 5,000 years ago, the current interglacial trend began to veer upward and away from the stacked average, which continued dropping. In the last 2,000 years, the methane concentration rose above the upper bound of the previous variations. This mismatch further supports the conclusion that natural explanations for the CH_4 increase of the last 5,000 years are falsified.

Observational evidence shows decreases in methane emissions during the last 5,000 years from two sources generally regarded as the primary natural ones (see Chapters 2 and 10). This evidence firmly rules out the possibility that a natural methane increase from northern hemisphere sources explains the CH_4 rise. As noted in Chapter 10, one model simulation suggests that methane emissions from a source in

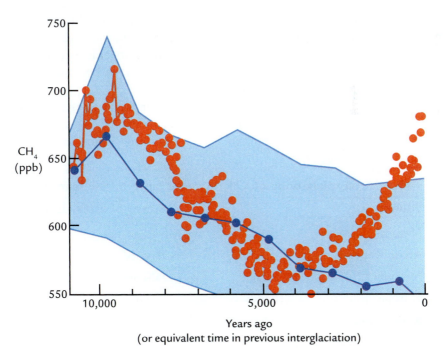

FIGURE 12-1 CH_4 trend during the current interglaciation (*red*) compared to the average (*dark blue*) and standard deviation (*light blue*) of previous interglaciations. *[CH_4 values from EPICA Community Members, "Eight Glacial Cycles from an Antarctic Ice Core," Nature 429 (2004): 623–628.]*

southern Amazonia grew strong enough during the last 5,000 years to play a significant role in the CH_4 increase, but other simulations indicate that natural variations in monsoonal regions of the Southern Hemisphere are secondary to the effects of larger changes in the Northern Hemisphere. This disagreement among models needs to be explored further.

The evidence of rising methane emissions from anthropogenic sources during the last 5,000 years is straightforward (Table 12-1). All four anthropogenic sources of methane—rice irrigation, livestock tending, biomass burning, and human waste—increased in strength as human populations grew and agriculture expanded (see Chapters 4, 5, 7, 8, and 10). The pioneering effort by Dorian Fuller and colleagues to quantify the expansion in total area irrigated for rice agriculture suggests that this source of CH_4 emissions can account for about 75% of the observed increase in atmospheric CH_4 concentration since 5,000 years ago. If future archeological investigations verify this claim, natural explanations for the CH_4 increase would be further refuted.

In addition, the pre-industrial spread of methane-emitting livestock mapped by Fuller and colleagues would have added to the anthropogenic CH_4 total, as would increases in emissions from increased biomass burning and human waste. At this point, anthropogenic explanations for the CH_4 increase during the last 5,000 years have become not just viable, but increasingly promising.

In summary, most of the evidence argues persuasively against natural explanations for the rise in CH_4 concentrations during the last 5,000 years, and comparisons to the downward trends in previous

TABLE 12-1 Explanations of CH_4 Increase Since 5,000 Years Ago

Sources	Status of explanation
Natural	
Arctic wetlands	falsified
N. tropical/subtropical wetlands	falsified
S. tropical/subtropical wetlands	viable (size uncertain)
Anthropogenic	
Rice irrigation	viable (large)
Domesticated livestock	viable (potentially large)
Biomass burning	viable (size uncertain)
Human waste	viable (small)

interglaciations meet a high standard of falsification. Future progress on this issue seems most likely to come from additional syntheses of archeological and historical data, and from modeling studies that use these data to make quantitative estimates of anthropogenic impacts on methane emissions.

Falsifying Explanations for the Rising CO_2 Trend

The question of the source (or sources) of the 22-ppm CO_2 increase during the last 7,000 years remains open to debate (Table 12-2). The two natural sources proposed—ocean carbonate compensation and the growth of coral reefs—are theoretically viable, but neither has been fully tested by field studies (see Chapter 11). In view of the enormous amount of field evidence that would be required to do so, it is not clear that such tests are even practical.

The 40-ppm CO_2 anomaly proposed in the early anthropogenic hypothesis is a combination of (1) the 22-ppm rise observed in ice-core trends, and (2) the natural downward CO_2 trend of approximately 18 ppm that the hypothesis says was nullified by anthropogenic emissions. Early modeling attempts to simulate pre-industrial land-use appeared to falsify the anthropogenic explanation because the small early populations existing thousands of years ago only seemed able to account for a CO_2 increase of 3–5 ppm. But a later simulation by Jed Kaplan and colleagues based on actual historical data documented higher early per capita land use and yielded a larger carbon emission estimate, one equivalent to about 24 ppm (see Chapters 8 and 11). In addition, a geochemical carbon-isotope index used to weigh all known inputs and outputs of terrestrial carbon during the last 7,000 years yields a similarly large 23-ppm estimate of direct anthropogenic

TABLE 12-2 Explanations of CO_2 Increase Since 7,000 Years Ago

Sources	Status of explanation
Natural	
Ocean carbonate compensation	viable (quantification difficult)
Growth of coral reefs	viable (quantification difficult)
Anthropogenic	
Deforestation	viable (potentially large)
Coal, peat burning, deep erosion	viable (small)
Ocean CO_2 feedback	viable (moderate in size)

emissions, if highly credible published estimates of large carbon burial in Arctic peatlands are used in the calculation.

Early burning of coal and peat may have added a few parts per million to the 23- to 24-ppm total from anthropogenic emissions, but the most promising additional source of CO_2 lies in feedback from warm oceans. Feedback resulting from reduced CO_2 solubility in warm seawater is estimated at around 8 ppm, which would bring the total anthropogenic CO_2 anomaly to about 33 ppm. The possibility of additional feedback from other ocean processes (particularly in the Antarctic region) will be evaluated with future models that couple ocean/atmosphere circulation and carbon processes. More work is also needed to refine all of the above estimates of anthropogenic emissions, particularly those from deforestation.

As was the case for methane, CO_2 trends during previous interglaciations can be thought of as natural experiments Nature has run on our behalf, providing an additional basis for choosing between the natural and anthropogenic hypotheses. The prevalence of downward

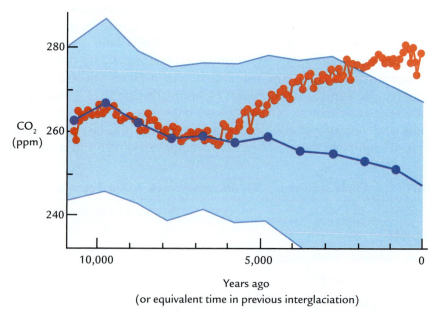

FIGURE 12-2 CO_2 trend during the current interglaciation (*red*) compared to the average (*dark blue*) and standard deviation (*light blue*) of previous interglaciations. [CO_2 values from EPICA Community Members, "Eight Glacial Cycles from an Antarctic Ice Core," Nature 429 (2004): 623–628.]

CO_2 trends over upward ones in six of seven test cases shows that Nature's verdict favors the anthropogenic explanation (see Chapter 9).

The wide variation in CO_2 trends among previous interglaciations (see Chapter 9, Figure 9-3) again makes comparisons somewhat difficult, but the CO_2 trend in the current interglaciation can be compared to an average of the previous ones (Figure 12-2). Prior to 7,000 years ago, the CO_2 trend during the current interglaciation coincided almost exactly with the average trend for the earlier interglaciations. This close agreement suggests that the CO_2/climate system prior to 7,000 years ago was behaving in a natural way. But after that date, the CO_2 trend began to veer upward from the average trend of the previous interglaciations and (like methane) rose above the one-standard-deviation range of previous variability nearly 2,000 years ago. This comparison further suggests that the upward CO_2 trend of the last 7,000 years is not natural.

Summary
Neither the natural nor the anthropogenic explanation for the rising CO_2 concentration during the last 7,000 years can at this point be falsified, but several lines of evidence have recently strengthened the case for the early anthropogenic hypothesis: land-use modeling, evidence of CO_2 feedback from warm oceans, and additional comparisons with CO_2 trends during previous interglaciations. Future progress in this debate could come from many areas, but increased coverage and analysis of archeological data to quantify land clearance and carbon emissions seems a particularly promising approach.

Additional Resources
Popper, K. R. *The Logic of Scientific Discovery*. New York: Routledge, 2002.
Ruddiman, W. F., J. E. Kutzbach, and S. J. Vavrus. "Can Natural or Anthropogenic Explanations of Late-Holocene CO_2 and CH_4 Increases Be Falsified?" *The Holocene* 25 (2011): 793–801. doi:10.1177/0959683610387172.

CHAPTER 13

PARADIGM SHIFTS

In his 1962 book *The Structure of Scientific Revolutions*, Thomas Kuhn largely sidestepped Karl Popper's idea that science proceeds in incremental stages by constant testing and falsification. Instead, Kuhn proposes that science moves forward in short spurts during which new ideas replace older ones, separated by longer intervals of time spent in a more quiescent or even sluggish mode he called "normal science."

During the intervals of normal science, an existing explanation for a particular set of phenomena works well enough for a long enough period of time that it becomes a widely held assumption, called a paradigm. Kuhn describes a paradigm as a combination of knowledge, shared general assumptions, conceptual models, schools of thought, and a "disciplinary matrix." In Kuhn's view, textbooks propagate existing paradigms to successive generations of students and future scientists by conveying a falsely tidy picture of the actual state of science. Often missing from textbooks is the sense of the underlying drama in the past when ideas were in conflict, and new paradigms were poised to emerge. Textbooks often emphasize the state of knowledge that exists today as if by looking through a rearview mirror, rather than through discussing the active process through which knowledge actually developed.

Kuhn dismisses most of the research carried out in the normal-science phase as "mop-up work," efforts that merely reinforce and refine an existing paradigm. Today's scientists often describe this kind of effort as "incremental research," which carries the negative connotation that this research is of lesser importance. Sometimes these contributions are derisively referred to as "LPUs" (least publishable units)—short, narrowly focused papers that report tiny advances and boost a scientist's apparent productivity. Kuhn claims that researchers conducting normal-science research usually do not aim to produce surprising new ideas, but simply elaborate on already accepted ones.

Some methods of normal science certainly seem less than exciting: increasing the accuracy or precision with which certain facts are known, or testing previous results by replicating the same experiment. Still, this work can result in basic scientific knowledge that is applied to practical scientific or engineering fields, and can be of enormous value to humanity. Efforts that devise instrumentation to extend the range of available measurements, or to measure previously unknown variables, can also be a very important and more creative feature of normal research.

Kuhn also notes a much more interesting aspect of science that attracts many young people—puzzle-solving. Some people love the challenge of grappling with and solving complicated puzzles. A related attraction for ambitious young scientists is the satisfaction of being the first person ever to figure out a puzzle, to solve a particular problem. The attraction of this kind of work comes from a combination of the pure joy of discovery and ego gratification.

Occasionally, the process of puzzle-solving leads to anomalous results that cannot be explained within the framework of the existing normal-science paradigm. Scientific investigators may become aware that something has "gone wrong" with an experiment or observation, that expectations based on the existing paradigm were not fulfilled. Kuhn points out that many of these discoveries come either from young scientists first entering a field, or from older ones who have recently switched to a new field. In both cases, Kuhn suggests, their minds are less clouded by preconceived opinions and are more open to fresh discoveries and new possibilities. Rather than rejecting contradictory new results, these scientists may abandon the process of normal science and search for the cause of the experimental "failure," thereby groping toward a new understanding of the issue. In time, this pursuit may lead to a new paradigm that does a much better job of explaining the newly observed and existing facts than the previous conceptual framework did.

Acceptance of New Paradigms

Wider acceptance of a new idea obviously depends not just on how well it accounts for otherwise unexplained phenomena, but also on its reception within the broader scientific community. According to Kuhn, many scientists feel a natural resistance toward novel ideas that disagree with deeply held convictions, prior expectations, and their own previous publications.

The amount of time that passes before a new paradigm gains wide acceptance can vary greatly. Individual scientists may experience a sudden, almost intuitive flash of realization that an old idea does not work and a new one does. The reasons why one particular scientist makes this conversion while another does not are as varied as the personalities and life histories of the scientists. Broader acceptance may come quickly if enough scientists perceive that the previous paradigm has failed, and the field is widely seen to have fallen into a state of crisis.

For example, geologists in the early 1900s had noticed the remarkable fit between the eastern coastline of South America and the western coastline of Africa, but no one had been able to devise a physically plausible explanation. Most scientists simply chose to ignore this striking piece of evidence and held to the paradigm that continents cannot move.

During the 1950s and 1960s, however, new marine geological and geophysical observations found a high-elevation ridge running down the middle of the Atlantic Ocean. It had a shape that matched both of the continental coastlines, and its position was marked by a series of earthquakes that indicated present-day tectonic activity. At this point, it became widely obvious that the continents had once been joined but had then split apart along the line marked by the mid-ocean ridge and were still moving apart today. This realization soon led to the plate-tectonic hypothesis (now theory).

Kuhn also notes examples in which new paradigms have been delayed for decades or longer. The scientific community may for a while attempt *ad hoc* modifications of the older paradigm in an attempt to resolve the emerging contradictions. Or a new paradigm may simply be ignored for a while, with much of the scientific community remaining in a state of *de facto* denial. In other cases, imperfections in the initial formulation of the new hypothesis may delay its acceptance until the shortcomings are addressed and incorporated in an improved version.

Often, older scientists will continue to reject a new paradigm even as many younger ones have become convinced by it. As the physicist Max Plank once said, "A new scientific truth does not triumph by convincing its opponents and making them see the light, but rather because its opponents eventually die and a new generation grows up that is familiar with it."

In his 1963 book, Kuhn summarized several of the largest paradigm shifts in scientific history: from the Earth-centered solar system of Ptolemy to the Sun-centered cosmos of Copernicus; from the belief that air contained a highly flammable material called phlogiston to the discovery of the chemical element oxygen; and from the classical mechanics of Newton to the relativistic universe of Einstein. Kuhn also noted that smaller revolutions can alter the views of members of professional specialties in a similar way to these large-scale shifts.

The early-anthropogenic hypothesis discussed here clearly ranks at a level of importance well below the examples cited above. Yet it does pose a direct challenge to our understanding of the climate in which human civilization developed, and its potential impact is not limited to a small scientific subspecialty. Although the hypothesis has not at this point become a new paradigm, evidence indicates (see Chapters 8–12) that it could become one in the future. What follows in the rest of this chapter is a personal summary of what I have seen about the process of science from my own experience in developing this hypothesis, and in working to resolve its shortcomings.

My Personal Journey Toward a Possible New Paradigm

In the scores of invited lectures on this subject I have given over the last ten years, I have often been asked how I came up with the early anthropogenic hypothesis. In a sense, this is actually a more general question about the inspiration for new ideas. In this case, the hypothesis started when I saw the first low-**resolution** ice-core records of methane trends spanning the current interglaciation. As explained in Chapter 2, the CH_4 increase during the last 5,000 years disagreed with my then-limited understanding of how natural sources of methane were changing during that time. From my interactions with John Kutzbach and Alayne Street in the **COHMAP Project** during the 1980s, I had learned that the large Asian and North African tropical monsoons had been progressively weakening for the last 10,000 years. I had also learned from ice-core literature that these northern-tropical wetlands filled by monsoon rains were considered the largest natural sources of methane on the planet.

Putting these two ideas together, I reasoned that the methane trend should have fallen during the last 10,000 years. And it did fall from 10,000 to 5,000 years ago, as expected. But then the methane concentrations reversed direction and unexpectedly headed upward. In my book *Plows, Plagues, and Petroleum*, I described this as my "Colombo moment," the feeling that "there's *just this one thing* that's bothering me," that sense that motivated the rumpled TV detective to keep poking at a crime-scene mystery. Recognizing this seemingly wrong-way methane trend was the first step in this long journey, and it was largely a result of my previous experience in the COHMAP group. The second step was the fact that I paid attention to this apparent anomaly, rather than ignoring it.

As a contributing factor, by mid-career I had become a climate-system generalist, after starting off initially as more of a specialist (in marine sediments). I had drifted in the generalist direction because I was more fascinated by the larger picture of how things fit together in the global climate system than I was by the aspects of data production, lab work, or technique innovation.

It also helped that I was writing a new college-level textbook (*Earth's Climate: Past and Future*) that covered the full spectrum of paleoclimate studies. Trying to do an informed and balanced job writing that book forced me to acquire a working knowledge in a wide range of areas, including some subjects I had not kept up with in this rapidly expanding field. Some research scientists become expert in one or two disciplines that span particular intervals of Earth's history, making a name for themselves within those areas, but they don't keep close track of other fields of science. Writing the first edition of the textbook in the late 1990s forced me to broaden my perspective well beyond fields that I knew from my own professional experience.

To some extent, I was also a newcomer to several fields that formed part of the hypothesis. I didn't work in ice-core studies, although I had kept track of the fascinating results emerging from that field because of their importance to the climate system. I was also almost entirely unaware of the vast range of research by archeologists and researchers in related disciplines on the origins and proliferation of early agriculture and land clearing. And I had only a passing familiarity with areas such as carbon cycling, carbon budgets, and the complexities of ocean carbonate chemistry. In a sense, this personal history fits Kuhn's notion that new ideas often come from people who are new to a field (although in this case definitely not from a young scientist!).

New hypotheses often follow one of two paths:

Hypothesis → Falsified by evidence → Discarded
Hypothesis → Supported by evidence → Theory (→ Paradigm?)

Most hypotheses are falsified because they fail to explain contradictory observations, and they are discarded and forgotten. A small number of hypotheses that are found to be in reasonable agreement with a wide range of evidence may gain enough support to be viewed as theories. Fewer still are sufficiently broad in scope that they overturn important conventional views and become new paradigms.

The computer scientist John Mashey, an astute observer of work in the field of climate science, has stated that he finds the early anthropogenic hypothesis a particularly vivid and interesting example of science in action. In 2005, he posted a comment on the blog realclimate.org that described four degrees of acceptance or rejection of new scientific ideas:

Case 1: The new idea goes nowhere.

Case 2: The new idea is refuted because stronger evidence favors competing explanations.

Case 3: The new idea makes claims based on valid supporting evidence, but it faces competition from other explanations that are based on comparably convincing evidence. As a result, the new hypothesis remains under consideration, but is not universally accepted.

Case 4: The new idea triumphs over competing explanations and is accepted as a theory.

As of 2005, Mashey placed the early anthropogenic hypothesis in Case 3. He felt that it had strong supporting evidence but faced competition from other hypotheses with comparably strong evidence pointing toward a natural origin for the greenhouse-gas rises. I found this assessment fair. The comparisons to gas trends in previous interglaciations supported the hypothesis (see Chapters 2 and 3), but other lines of evidence summarized in Part 3 (see Chapters 8 and 9) argued against it at that time.

After the hypothesis was published late in 2003, I began trying to convince others that it had merit, while also exploring the criticisms other scientists had raised. My early efforts to have follow-up papers on my hypothesis published in prominent scientific journals were frustrating. I submitted three papers to the journal *Nature*. All of them were sent out for peer review, and all were rejected because they received a mix of reviews: one mildly positive, one more neutral, and one negative. In the intense competition to have papers published in *Nature*, one negative review is usually enough to result in rejection. This basic fact of scientific life turned out to be particularly hard on a new hypothesis that was challenging the existing wisdom.

I also submitted two manuscripts to the prominent journal *Science*, where the results were perplexing. The editor responsible for my manuscripts didn't even send them out for official reviews. Instead, they were returned with a standard form-letter rejection stating that the paper was not of broad enough interest to merit publication in *Science*, and that it belonged in a specialty journal. I thought this reply was nonsense. The respected veteran *Science* writer Dick Kerr had by then found my hypothesis interesting enough to write the first of two articles on it, and the book review section of *Science* had carried a positive review of my 2005 book *Plows, Plagues, and Petroleum* (which featured the anthropogenic hypothesis prominently). Apparently, my work was of "broad enough interest" to merit press coverage and book reviews in *Science,* but not important enough for my papers to even be sent out for review by researchers in the field.

As for the supposed "lack of broad interest," I had for years been giving a dozen or more invited lectures a year. Most of the invitations had come from universities and colleges, and many because of input from student committees. Faculty members often told me that the students expressed interest in seeing new ideas challenge old ones, and that they found science in action more exciting than listening to the received wisdom on issues settled long ago. Their reactions add an interesting dimension to Kuhn's statements about the sluggishness of "normal science." Many students also seemed to enjoy the blending of science and the humanities (especially history) that is present in the early anthropogenic hypothesis.

As I gave all those lectures, I found it hard to gauge the true range of opinions on the early anthropogenic hypothesis. My best guess was that a sizeable minority of scientists believed the hypothesis, a similarly sized minority did not, and a large "silent majority" remained uncommitted—intrigued by it, and still open-minded, but far from convinced.

Resistance to the hypothesis came in several forms, one of which is direct challenges from scientists who had found what appeared to be contradictory evidence. Part 3 covered several prominent examples: land-use modelers whose simulations suggested that early deforestation was not that extensive (see Chapter 8), scientists who questioned the choice of how to align gas trends in previous interglaciations with those in the current one (see Chapter 9), and carbon geochemists who cited carbon-isotopic evidence that initially seemed to preclude large emissions of terrestrial carbon since 7,000 years ago (see Chapter 11). These challenges, coming from top-caliber scientists, were thoughtful and plausible. As described in Part 3, the criticisms motivated me and other scientists to look more carefully at several issues. This process of testing, questioning, and criticizing is how science should work, and in most cases how it does. Challenges like these are well worth taking seriously, because any ideas with real merit will only be improved by careful examination.

In one case, resistance of a very different kind came from a highly regarded paleoclimate scientist who did not accept my hypothesis and made several disparaging remarks to reporters who were looking for comments on it. He was quoted as calling my ideas "a bunch of bosh," "an insane argument," "total and utter nonsense," and "bad science."

Because scientists have a wide range of personalities and styles of expression, varying degrees of intellectual and verbal opposition to new ideas are to be expected. But the disparaging tone of these remarks was unusual, especially coming from such a distinguished scientist. In any case, this criticism just made me all the more determined to pursue the scientific questions the hypothesis had raised.

Much more typical of the scientific community at large was a kind of inertia-based resistance along the lines that Kuhn described in his book. I can best summarize it by this imagined sequence of reactions to the hypothesis from a "typical" scientist:

> *This new idea disagrees with what I've been taught.*
> *It involves issues that lie somewhat outside the field I know best.*
> *I don't have time to look into it closely, so I'll just lean on the experts.*
> *A few very prominent experts don't believe it, so I'll trust them.*

Few scientists have the time to dig into and evaluate the detailed evidence behind an argument, especially if it lies on the margins of their field or beyond it (as, inevitably, one or another aspect of my hypothesis does). Given this attitude, and the tendency to rely on expert opinion, some scientists have no doubt been influenced by the intemperate criticism and have held to the existing paradigm despite growing evidence against it.

I also suspect that a more subtle kind of unconscious resistance could be coming from some scientists studying pre-industrial climatic and environmental changes. Several government-funded programs have for years supported efforts to document pre-1800 changes in detail in order to define the natural climate state (the baseline) against which the accelerating disturbances of the Industrial Era can be compared. Yet if the early anthropogenic hypothesis is correct, those research programs are based at least in part on a false premise. Climate has not been "natural" for the last 7,000 years; instead, it has become increasingly anthropogenic over time.

In any case, the decade-long stand-off between the natural and anthropogenic hypotheses now shows some signs of coming to an end. Since 2009, new studies have begun bolstering the case for the early anthropogenic hypothesis, culminating in a 2011 special issue of the journal *The Holocene* on this topic (I was one of three editors for this issue). Everyone who had published papers on this debate was invited to participate in the issue, although several prominent critics of the hypothesis declined. Nevertheless, the issue still included papers from both sides of the debate.

Several of the papers in this journal reported new evidence suggesting that a new paradigm may soon emerge. The original argument that the gas trends of the last several thousand years went the wrong way (see Chapters 2 and 3) was upheld by the closer inspection described in Chapters 9, 10, and 11. Newer research showed that those upward gas trends cannot be natural. In addition, historical and archeological evidence has invalidated previous land-use reconstructions based on the assumption of small and constant per capita land clearance (see Chapter 8). Instead, newer simulations point to much larger early land clearance and cultivation. The size of the pre-industrial CO_2 and CH_4 emissions estimated from these new studies is more in line with the early anthropogenic hypothesis, although additional factors are still required to account for the size of the CO_2 rise between 7,000 and about 4,000 years ago (see Chapter 11, Figure 11-9). Finally, closer inspection has revealed that the carbon-isotope index does not limit pre-industrial CO_2 emissions to a small size. Large carbon burial in Arctic peat deposits during the last 7,000 years requires large anthropogenic emissions (see Chapter 11). In summary, the evidence favoring the early anthropogenic hypothesis still stands or has even been strengthened, while much of the evidence that once seemed to contradict it now appears to be flawed or at least weakened.

I have at times been asked: "Why couldn't the CO_2 and CH_4 increases of the last few thousand years have been caused by some natural process that we just haven't been smart enough to think of yet?" Aside from the growing field evidence from history and archeology indicating that the gas increases were likely anthropogenic, the comparison to previous interglaciations provides a "universal" answer to this question. Because the natural CO_2 and CH_4 trends early in those prior interglaciations were downward, they categorically refute *any possible* natural explanation for the upward trends of the last few thousand years, including explanations no one has thought of yet.

Finally, my ongoing explorations of this issue have provided me with two unanticipated personal insights.

One unexpected scientific lesson is how important it has been to have had from the outset a strongly held point of view based on solid evidence that made me more willing to challenge other evidence in various fields. Specifically, the conclusion that I drew early on from the mismatch between the CO_2 and CH_4 trends during the previous interglaciations and those in the current one helped me hold firm against strong criticisms of various other aspects of the hypothesis. I felt that those criticisms must somehow be flawed, but for a long time I didn't know how. Eventually, those flaws emerged. Without the evidence from the wrong-way gas trends, I might have given up the debate some time ago.

A more sociological lesson that I learned is the need to stay open to criticism. Calm acceptance of criticism is not encoded in our DNA; reacting to people as either friend or foe is much more instinctive. In science, this trait carries dangers for those with naturally confrontational personalities, and for some people it can become a bigger problem in old age if they become more inflexible, as some do. In my case, when the worst criticisms came, I did my fair share of wincing and short-term denial, but I bounced back, kept an open mind, and tried not to make the debate personal. By now I think I could almost write a book on the sociology of science, and the psychology of scientists!

Summary
My general impression, after a decade of involvement in this contentious issue, is that the process of science works well most of the time, although not always in a straight line. Often it progresses in a zigzag, and sometimes it even runs backward for a while. Passions flare, but evidence and logic eventually win out, as I think they are doing in this case, as the evidence builds. The early anthropogenic hypothesis is alive and well, and gaining ground.

Additional Resources
Kuhn, T. S. *The Structure of Scientific Revolutions.* 3rd ed. Chicago: University of Chicago Press, 1996.

Ruddiman, W. F. *Plows, Plagues, and Petroleum.* Princeton: Princeton University Press, 2005.

Weart, S. *The Discovery of Global Warming.* 2nd ed. Boston: Harvard University Press, 2008. http://aip.org/history/climate/index.htm.

CHAPTER 14

AN EMERGING PARADIGM FOR THE ANTHROPOGENIC ERA?

Near the turn of the millennium, the aquatic ecologist Eugene Stoermer began using a new term, the **Anthropocene**, to refer to the period in which human influences on Earth's environment and its climate outweighed natural changes. In 2002, the atmospheric chemist and Nobel prize winner Paul Crutzen joined with Stoermer to publish a short article that formalized this new view and proposed that the Anthropocene started in the early to middle 1800s. Their proposal quickly gained favor with many climate scientists who saw ample evidence that major human interference in the climate system began with the initial rise of greenhouse-gas concentrations during the Industrial Era around 1850 and has grown exponentially since then. Without question, the last century and a half has seen increases in the human footprint on this planet that are without precedent (Table 14-1).

During this same interval, many innovations have benefited humanity by making labor less arduous, by improving human health and food supplies, and by lengthening average life spans. As a result, the planet's population has surged from around 1 billion people in 1850 to over 7 billion today. Many people still lack modern comforts or even life's basic necessities, but far more people than before (and a higher percentage of the world's population than previously) now have their basic necessities

TABLE 14-1 **Major Transformations During the Industrial Era**

Innovation	Environmental consequences
Mechanized agriculture	Farming in prairies and steppes
	Deep soil erosion in some areas
Damming of rivers	Siltation of catchments behind dams
Mechanized earth-moving	Greater than all similar natural processes
	Ecosystem fragmentation and species loss
Industrial (nitrogen-rich) fertilizers	Runoff to coasts, algal blooms
Industrial power plants	Release of sulfates to atmosphere
	Acidification of lakes, streams, rivers
Use of fossil fuels	CO_2 emissions
	30% increase in ocean acidity
Use of CFCs for refrigeration	Creation of ozone hole
Mechanized fishing methods	Depletion/collapse of primary fisheries
Mechanized irrigation	Human use of >50% of river water
	Pollution of many rivers
Mechanized pumps	Depletion of groundwater

met and enjoy a moderate level of comfort. From a larger perspective, however, many of these improvements have come at a substantial cost to the environment, including the changes listed in Table 14-1. These negative impacts are part of Crutzen and Stoermer's concept of the Anthropocene.

In a larger sense, the Industrial Era view of the Anthropocene concept reflects a paradigm about human effects on climate that has prevailed for decades. This view acknowledges earlier (pre-industrial) effects, but sees them as much smaller than those that have occurred during the Industrial Era.

But is most of the history of our impact on this planet really limited to just the last century and a half? Parts 2 and 3 of this book have suggested that the answer to this question is no. Archeological and historical evidence show that human actions transformed much of Earth's surface long before the start of the Industrial Era.

Recent evidence suggests that a new paradigm may be just about to emerge—one with much earlier and much larger human impacts than those implicit in the paradigm of an Industrial Era Anthropocene.

If so, what would this new paradigm look like? This chapter answers this question by examining three measures of large-scale human influence: land use (mainly forest clearance), greenhouse-gas emissions, and resulting effects on global temperature.

Anthropogenic Forest Clearance

By far the largest transformation of Earth's surface by humans has been agriculture. Millions of square kilometers of croplands and pastures have replaced areas that were previously natural forests and grasslands. All of our buildings, roads, parking lots, and other built structures don't even come close to the area that human civilization has devoted to agriculture.

The paradigm of an Industrial Era Anthropocene emphasizes rapid deforestation since 1850, compared with much slower rates and cumulatively smaller amounts of clearance before that time. Several papers in the last two decades have estimated that almost two-thirds of the total deforestation by humans has occurred since 1850, and that much of the earlier clearance happened during the centuries just prior to the Industrial Era (Figure 14-1A). This view seems to be supported by evidence for heavy clearance in recent decades across Amazonia and other parts of Central and South America.

In contrast, the early anthropogenic view proposed in 2003 suggests just the opposite: that well over half the total clearance had already occurred before 1850 (Figure 14-1B). In this view, the slow but cumulatively large clearance in pre-industrial times exceeded the explosive but brief clearance during the last century and a half—in effect, a slow pre-industrial turtle outrunning the swift industrial hare. This view particularly focuses on evidence for large-scale early clearing of forests in the Old World (chiefly Eurasia).

The proposed Industrial Era Anthropocene paradigm is vulnerable to several criticisms. First, the actual database of fully reliable land-use surveys in most countries only extends back a few decades (Figure 14-2). Second, many of the surveys (and the less complete information that preceded them) show *re*forestation in recent decades, not *de*forestation. France, Britain, Sweden, and other European countries with longer records had already begun reforesting by 1850 and have continued to do so during the rest of the Industrial Era. As noted in Chapter 8, Alexander Mather called this reversal from the long previous deforestation phase to the recent interval of reforestation the *forest transition.*

The same reversal has occurred in China since the 1970s, in large part because of directives from the central government to plant trees as

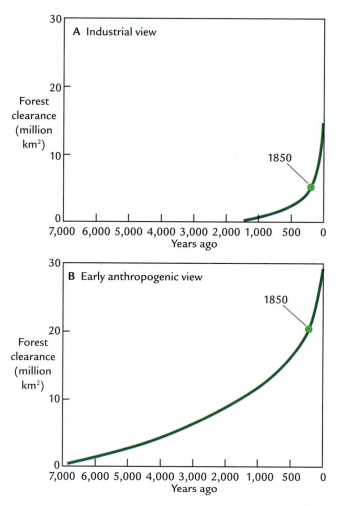

FIGURE 14-1 Two highly schematic views of the history of cumulative forest clearance: (A) mainly Industrial Era clearance; (B) mainly pre-industrial clearance. (Note scale change at 1,000 years ago.)

windbreaks and plantations. Reforestation has also occurred in Russia since the 1920s, due partly to the loss of 20 million people during two world wars, forced relocations of ethnic populations, and a net loss of population in recent decades. Even east-central North America, which had been rapidly deforested during the eighteenth and nineteenth centuries, was already reforesting by 1900. In many regions, reforestation occurred when mechanization made larger-scale, more cost-effective agriculture possible on fertile prairie and steppe soils.

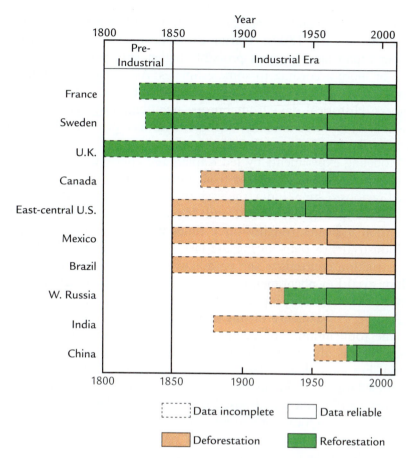

FIGURE 14-2 Lengths of reliable and less complete land-use records just prior to, and during, the Industrial Era, including changes from deforestation to reforestation.

The limitations of these short land-use records cast doubt on the reliability of methods that use population trends to estimate past land use, the approach used by most studies that take the Industrial Era Anthropocene view. How can past *de*forestation during times of smaller populations be estimated from recent *re*forestation during times of much larger populations?

Another major vulnerability of the Industrial Era Anthropocene view is its reliance on the assumption of small, constant per capita clearance to project back in time to estimate past deforestation. A range of evidence on land use (see Chapter 8) provides a completely different picture. Historical evidence shows that farmers 2,000 years ago were using four to five times as much land per capita as those in late pre-industrial times (the 1700s to early 1800s). During the millennia prior to the historical era, per capita land use was probably higher still, perhaps by a factor of 10 or more according to some estimates. As a result, the amount of forest cleared by these early farmers was far out of scale to their modest numbers. Historical data from Europe unambiguously support simulations that show large amounts of forest clearance prior to the Industrial Era (see Chapter 8, Figure 8-6).

Modeling experiments by Jed Kaplan and colleagues have projected the historical land-use patterns from Europe across other heavily populated regions: China, India, the eastern Mediterranean, parts of Southeast Asia and tropical Africa, and Mexico. On a global basis, their simulation indicates roughly three times more forest clearance before the Industrial Era than during the last 150 years (Figure 14-1B). This reconstruction contradicts the paradigm that deforestation was mainly an Industrial Era phenomenon (Figure 14-1A), but it matches the basic claim of the early anthropogenic hypothesis.

Another consequence of this reanalysis by Kaplan and colleagues is that the total amount of deforestation from 7,000 years ago to the present day is about two times the total amount proposed in the Industrial Era view (Figure 14-1). The Industrial Era Anthropocene view had been largely based on the assumption that the amount of forest clearance can be calculated based on the present-day amount of cultivated land in regions with ample precipitation. But this assumption failed to take into account areas that had once been deforested, and then were later degraded to the point where agriculture was not possible or land was simply abandoned and left fallow. Including these once-forested, but not currently cultivated, regions doubles the total amount of deforested land.

Ecologists often analyze large-scale vegetation in terms of **biomes**, groups of ecosystems (including certain vegetation types) that occupy broad areas in response to patterns of temperature and precipitation. For example, high-latitude evergreen forests are dominated by spruce, fir, and larch trees that can withstand very cold temperatures, while mid-latitude deciduous forests feature trees such as oak, linden, maple, and hickory that are adapted to more temperate conditions.

The environmental scientist Erle Ellis takes the view that almost none of the vegetation on Earth today is natural, nor has it been for centuries or even millennia: ". . . nature is now human nature; there is no more wild nature to be found . . ." He examines vegetation in terms of **anthromes**, areas that have been transformed in varying degrees by humans. His analysis includes areas where biomes have been completely altered, such as cities or intensively cultivated farmland, but he also includes regions where the transformations are relatively subtle and incomplete, such as areas of shifting cultivation that are burned, cultivated for a few years, abandoned, and then perhaps cultivated again many years later after some of the vegetation has grown back. In addition, Ellis notes dwindling areas that are listed today as "forest" but have long since lost a substantial part of their carbon stock as nearby humans have gathered wood for heating, cooking, and construction of dwellings. Ellis' anthrome mapping approach further supports the idea that the start of major clearance occurred well before the Industrial Revolution.

This emerging reassessment of the long history of human land use indicates that the Crutzen-Stoermer paradigm of an 1850 start to major land clearance and use is incorrect (Figure 14-1). The large amount of pre-industrial clearance simulated by Kaplan and colleagues remains to be tested by more fieldwork and further modeling simulations, but the evidence is now sufficient to claim that much more land was cleared long ago than allowed for in the Industrial Era Anthropocene view. The alteration of Earth's surface began at small levels with the origin of agriculture more than 10,000 years ago, and it grew continuously through the large-scale clearance prior to the Industrial Era, without any obvious threshold change from a time of little alteration to an era of major changes. In summary, the emerging historical and archeological evidence discussed in previous chapters points to a new paradigm of the history of land use: humans cleared far more forest prior to 1850 than we have since then.

Anthropogenic Greenhouse-Gas Emissions and Atmospheric Concentrations

Most of the focus on anthropogenic greenhouse-gas emissions is on carbon dioxide because of its larger climatic effect (as a whole) relative to methane. The current Industrial Era Anthropocene paradigm focuses on the observed CO_2 increase of slightly more than 100 ppm since 1850 (Figure 14-3A). According to this paradigm, this CO_2 increase far exceeds the amount produced by the limited pre-industrial deforestation assumed in Figure 14-1A. CO_2 simulations based on such limited clearance suggest pre-industrial carbon emissions of 50 to 90 billion tons, equivalent to a CO_2 increase in the range of 3 to 6 ppm.

In contrast, the early anthropogenic paradigm proposes much larger pre-industrial gas releases. A crude estimate in my original 2003 paper

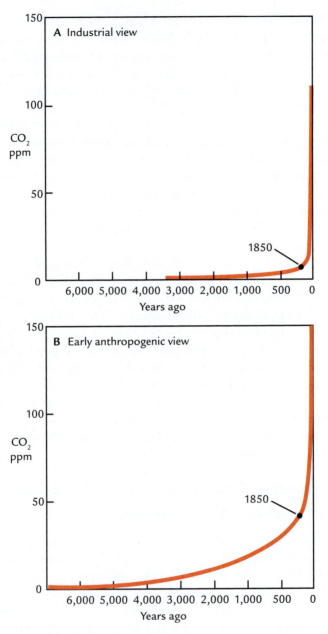

FIGURE 14-3 Two highly schematic views of the history of anthropogenic CO_2 increases: (A) mainly Industrial Era increases; (B) large pre-industrial increases. (Note scale change at 1,000 years ago.)

that introduced the early anthropogenic hypothesis suggested a cumulative pre-industrial release of 300 billion tons of carbon, roughly three to six times as much as proposed in the Industrial Era Anthropocene view. The hypothesis also suggested that the full anthropogenic CO_2 anomaly in pre-industrial times was 40 ppm, compared to just 3–6 ppm in the Industrial Era paradigm.

Subsequently, the extensive pre-industrial forest clearance simulated by Kaplan and colleagues yielded an estimated cumulative pre-industrial release of 340 billion tons of carbon. These emissions, along with much smaller amounts from early burning of coal and peat, would have boosted pre-industrial atmospheric CO_2 concentration by about 25 ppm, many times larger than the amount in the Industrial Era Anthropocene paradigm. CO_2 feedbacks could have brought the total anthropogenic effect to more than 30 ppm, close to the 40-ppm level originally posited in the early anthropogenic hypothesis.

The spread of rice irrigation, mapped by Dorian Fuller and colleagues, indicates that early anthropogenic emissions of methane from rice paddies can explain much of the roughly 100-ppb rise in atmospheric methane concentration between 5,000 and 1,000 years ago (see Chapter 10, Figure 10-4). Historical data showing larger per capita rice holdings 1,000 years ago provide additional evidence of early CH_4 emissions considerably larger than in the Industrial Era Anthropocene view.

In any case, because the available evidence is now more than sufficient to invalidate the conventional view that pre-industrial CO_2 and CH_4 emissions were negligible, the Industrial Era Anthropocene paradigm (Figure 14-3A) is giving way to a different view (Figure 14-3B). If future simulations based on historical and archeological evidence continue to favor gas emissions of the sizes proposed in the early anthropogenic hypothesis, the new paradigm would also mean that human effects on greenhouse-gas concentrations in the atmosphere started from a lower natural baseline than had been previously thought and have thus been larger in a cumulative sense (Figure 14-3B).

Since 1850, the CO_2 concentration in the atmosphere has risen from about 285 ppm to over 390 ppm, an increase of more than 37%. But if the late pre-industrial CO_2 concentration had fallen naturally to somewhere between 240 and 245 ppm, as predicted by the early anthropogenic hypothesis, the true anthropogenic CO_2 rise would have started at that lower baseline on its rise to more than 390 ppm, for a total anthropogenic increase of more than 60%.

Similarly, the CH_4 concentration has risen from 795 ppb to over 1,750 ppb during the Industrial Era, an increase of more than 120%.

But if the CH_4 concentration had fallen naturally to the predicted value of 450 ppb, the total anthropogenic CH_4 rise would have been from 450 ppb to more than 1,750 ppb, an increase of almost 190%. If confirmed, these revisions would constitute a new paradigm on the size and timing of human impact on greenhouse-gas concentrations.

Anthropogenic Effects on Climate

In the Industrial Era Anthropocene view, the large increases in greenhouse-gas emissions and atmospheric concentrations observed since 1850 have accounted for much of the total human effect on global temperature (Figure 14-4A). Standard estimates of Earth's temperature sensitivity to greenhouse-gas changes indicate that those well-documented increases have the potential to have warmed Earth's climate by about 2°C by now, ignoring complications from other factors.

But two other factors have kept the full Industrial Era warming from happening. First, **aerosols** (very small particles, mostly dust), generated by human activities such as industrial refining and production, have blocked incoming solar radiation and caused an opposing cooling, the size of which is not well understood. Second, the climate system takes decades to respond to any kind of climatic push, mainly because of the slow response of the upper layers of the ocean that cover 70% of Earth's surface.

As a result, the pulse of greenhouse-gas emissions emitted during the Industrial Era (and especially the large amount emitted in the last several decades) has been so rapid and so recent that the climate system has not yet had time to register its full warming response. Because of these two factors, the current net response to Industrial Era activities has been an observed increase of only about 0.8°C.

Counterintuitively, the relatively modest pre-industrial increases in greenhouse gases proposed in the early anthropogenic hypothesis could have caused a warming larger than that during the Industrial Era (Figure 14-4B). Even though the proposed anthropogenic CO_2 and CH_4 increases during pre-industrial times were smaller than those during the Industrial Era (Figure 14-3B), the climate system has had time to register them fully. Most of the delay imposed by the slow response of the ocean is measured in decades, but CO_2 and CH_4 concentrations rose gradually over many millennia, allowing time for the slow ocean response to keep pace.

For the proposed early anthropogenic anomalies (40 ppm for CO_2 and 250 to 350 ppb for CH_4), recent modeling experiments suggest

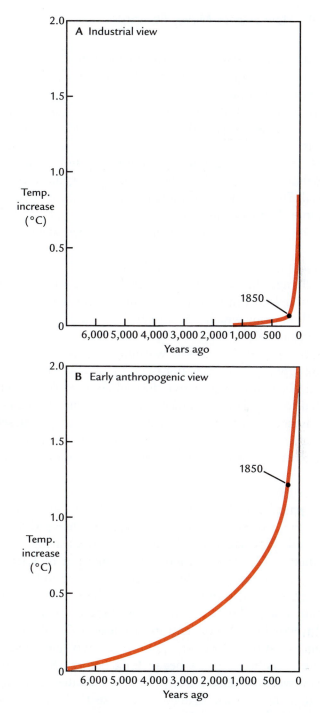

FIGURE 14-4 Two highly schematic views of the history of anthropogenic global temperature increases: (A) mainly Industrial Era warming; (B) larger pre-industrial warming. (Note scale change at 1,000 years ago.)

that the resulting global warming would have been 1.2 to 1.4°C. This amount is smaller than the potential (future) 2°C climate-system response to Industrial Era greenhouse gases now in the atmosphere, but it is larger than the net 0.8°C response registered so far to the combined warming from industrial greenhouse gases and the cooling from industrial aerosols.

If the pre-industrial greenhouse-gas releases from farming have caused a warming that was larger than the net anthropogenic warming during the Industrial Era, the effect of humans in warming this planet to date would be more than twice as large as the amount assumed in the current paradigm. Our total effect in warming the planet would be 2°C, not 0.8°C.

This new paradigm of early anthropogenic temperature impacts would not change the way scientists think about temperature changes during the Industrial Era and into the future. But it would redefine the natural baseline from which net human influences on temperature are measured (Figure 14-4).

One caveat is needed here—the warming caused by early greenhouse-gas emissions from farming was not the only way in which pre-industrial humans could have influenced pre-industrial climate. Two other potential effects related to early land clearance will be examined in the next part of this book: changes in Earth's surface **albedo** (reflectivity) caused by converting forests to croplands and pastures, and increases in aerosols (dust) in the atmosphere resulting from clearance, plowing, and other activities.

Summary

The current paradigm of human effects on planet Earth views 1850 as the start of the "Anthropocene," but recent evidence shows that this choice overlooks key aspects of human history. Clearing land for agriculture is the single largest human transformation of Earth's surface, and more areas of forest were cleared before 1850 than have been since. Greenhouse-gas concentrations in the atmosphere are a key component of the climate system, but if humans caused the reversals of the natural downward trends in CO_2 concentrations close to 7,000 years ago, and in CH_4 concentrations close to 5,000 years ago, global-scale human influences on greenhouse gases began long before 1850. Finally, if the pre-industrial warming effect from human-generated greenhouse gases is confirmed to be at least equivalent in size to, and probably larger than, the observed Industrial Era warming that has occurred to date, then the current paradigm of a young (post-1850) Anthropocene is also overlooking an important part of Earth's climate history.

A sensible solution to this issue would be to acknowledge that the Anthropocene actually developed in two stages: (1) an earlier phase of land clearance and increasing gas emissions that started at very small size but gradually grew much larger over 7,000 years; and (2) a later phase that has seen additional clearance, very rapidly increasing gas emissions, and an expanding list of other environmental impacts since 1850.

Additional Resources

Crutzen, P. J., and E. F. Stoermer. "The 'Anthropocene'." *Global Change Newsletter* 41 (2000): 12–14.

Ellis, E. C. "Anthropogenic Transformations of the Terrestrial Biosphere." *Philosophical Transactions of the Royal Society A* 369 (2011): 1010–1035. doi:10.1098/rsta.2010.0331.

Hooke, R. L. "On the History of Humans as Geomorphic Agents." *Geology* 28 (2000): 843–846.

Mather, A. S. *Global Forest Resources*. London: Bellhaven Press, 1990.

Steffen, W., P. J. Crutzen, and J. R. McNeill. "The Anthropocene: Are Humans Now Overwhelming the Great Forces of Nature?" *AMBIO 36* (2007): 1317–1321.

Part 4 Summary

The evidence summarized in Parts 2 and 3 of this book is sufficient to falsify some aspects of natural hypotheses proposed to explain the greenhouse-gas increases of the last several millennia. The absence of upward CH_4 trends during all seven previous interglaciations convincingly falsifies natural explanations for the rising methane trend during the last 5,000 years. The downward CO_2 trends in six of seven previous interglaciations comes very close to falsifying natural explanations for the CO_2 increase during the last 7,000 years, as does the departure of the recent CO_2 trend from the long-term average and its escape from the bounds of previous variations during the last 2,000 years.

Historical data from Europe and China, along with archeological data from southern Asia, reveal that per capita land use by farmers millennia ago was much higher than in recent centuries. This evidence falsifies the conclusion drawn from previous land-use modeling studies that pre-industrial forest clearance and carbon emissions were small. Higher early per capita land use means larger early CO_2 emissions, although research is still needed to quantify the amounts emitted. Estimates of CH_4 emissions from the gradual spread of rice paddies across southern Asia nearly falsify natural explanations for the methane increase of the last 5,000 years, and future estimations of emissions from the spread of domesticated livestock will inevitably further strengthen this conclusion.

The paradigm that large-scale human impacts on this planet have occurred mainly in the Industrial Era (the last 200 years or less) is no longer tenable. Forest clearance on most continents occurred mainly in pre-industrial millennia, along with substantial emissions of CO_2 and CH_4. Pre-industrial anthropogenic warming appears to exceed the temperature increase that has occurred during the Industrial Era. Based on emerging evidence, a new paradigm of very large pre-industrial anthropogenic influences is slowly coming into view.

Early Human Effects on Climate

PART 5

Chapter 14 described the potential emergence of a new paradigm in which climate would have become cooler during the last several thousand years had it not been for greenhouse-gas emissions from early agriculture. Climate-model simulations indicate that preindustrial cooling would have been amplified by snow and sea ice feedback at high northern latitudes by least 2°C in summer and as much as 4° to 5°C in winter. Chapter 15 explores the surprising claim from the early anthropogenic hypothesis that this cooling would have been large enough to allow ice sheets to begin forming in high northern latitudes.

Greenhouse-gas concentrations were not the only way that early farming altered climate. Cleared land surfaces responded to incoming solar radiation in ways that differed from their forested predecessors. Normal winter cooling at high latitudes is amplified in regions where dark evergreen spruce or fir forests that absorb solar radiation are converted to lighter croplands and pastures that reflect more solar radiation, especially when the cleared land is covered by snow. In contrast, clearance at snow-free lower and middle latitudes has little such effect, and may even cause local warming in summer. Chapter 16 investigates the climatic effect of the changes in land surfaces caused by forest clearance and assesses how much of the warming effect from greenhouse gases would be offset. Chapter 16 also explores whether or not dust particles raised by premechanized plows and other early farming activities could have blocked enough incoming solar radiation to cool climate.

Today, greenhouse-gas concentrations are rising rapidly, and are projected to do so far into the future. Because 10% to 15% of the CO_2 we emit into the atmosphere will stay there for tens of thousands of years, new ice sheets will not appear again on North America or northern Eurasia until the very distant future, if at all. An examination of the evidence in Chapter 17 concludes that Earth's long (2.75-million-year) history of ice-age cycles in the Northern Hemisphere has come to an end because of human activities, both pre-industrial and modern.

The major concern for the more immediate future is not about ice growing, but about ice melting. What fraction of the ice sheets now present on Antarctica and Greenland will disappear as greenhouse-gas concentrations rise and climate continues to warm? Eventually, tens to hundreds of millennia from now, the huge pulse of fossil-fuel CO_2 we have put in the atmosphere will be absorbed in the ocean, but during the next few centuries, greenhouse-gas concentrations will rise,

Earth's temperature will warm, and ice will melt. A looming question for humanity will be whether or not to intervene in the climate system by attempting to reduce the greenhouse-gas warming we are now causing.

Additional Resources

Ruddiman, W. F. "How Did Humans First Alter Global Climate?" *Scientific American* (March 2005), 46–53.

CHAPTER 15

IS THE NEXT GLACIATION OVERDUE?

In 1976, the marine geoscientists Jim Hays, John Imbrie, and Nick Shackleton confirmed an earlier hypothesis of Milutin Milankovitch that called on gradual changes in Earth's orbit over tens of thousands of years as the major influence on cycles of glaciation in the Northern Hemisphere. As explained in Chapter 1, the central argument of the Milankovitch hypothesis states that the strength of incoming summer radiation at high northern latitudes determines whether ice accumulates or melts. Aware of this confirmation of the hypothesis, and also of the 8% decrease in summer insolation at high northern latitudes during the last 10,000 years, many scientists in the late 1970s concluded that the start of the next northern hemisphere glaciation would occur in the not-too-distant future: perhaps as soon as in one or two thousand years.

From the late 1940s through the 1970s, global temperature (measured by thermometers) had been drifting erratically toward slightly cooler values. This trend, combined with the confirmation of the Milankovitch theory, persuaded a very small number of climate scientists to jump to the provocative conclusion that we might be heading into the next glaciation in the very near future. Nevertheless, most scientists took the more cautious (and reasonable) view that the start of a new glaciation could not be predicted with any precision based on the extremely slow pace of the orbital changes.

By the 1980s, measurements of atmospheric CO_2 pioneered by the atmospheric chemist Charles Keeling in 1957 showed a slow but persistent increase in CO_2 concentration. Given the known greenhouse effect of carbon dioxide, most climate scientists soon realized that, for the foreseeable future, the warming effect from the increasing CO_2 levels would overwhelm the much slower cooling effect from decreasing summer solar radiation. As it turned out, in 1980 global temperature soon resumed a general warming trend that had been underway since the late 1800s, and this rapid rise has continued to the present day.

The early anthropogenic hypothesis also casts light on the issue of glacial inception—the onset of the next glaciation. In its initial (2003) formulation, the hypothesis carried an important corollary conclusion: the next ice age should have begun a few thousand years ago, rather than happening centuries or thousands of years in the future. In this view, a new glaciation may actually be *overdue*, at least in the sense of new ice sheets growing in sparsely populated areas far to the north.

Although this claim may sound highly implausible, consider the evidence already presented in previous chapters about two key factors. First, summer insolation levels in the Northern Hemisphere have been decreasing for 10,000 years and are now just below the long-term average of the last several hundred thousand years (Figure 15-1). Compared to the natural long-term range, northern summer insolation has now dropped about 55% of the way from the high-end extreme toward the low end. Because ice sheets have been present for more than 90% of the last several hundred thousand years, the current values seem likely to be near or at a level favorable to a new glaciation.

The other key factor controlling advances and retreats of ice sheets is the concentration of greenhouse gases in the atmosphere, particularly CO_2. According to the early anthropogenic hypothesis, the reason a new ice age has not yet begun is the wrong-way CO_2 and CH_4 trends, both of which turned upward a few thousand years ago when they should have continued down (recall Chapter 12, Figures 12-1 and 12-2). Just before the Industrial Era, the gas concentrations reached 280 to 285 ppm for carbon dioxide and 790 ppb for methane, but the anthropogenic hypothesis claimed that the natural trend should have fallen to values of 240 to 245 ppm for CO_2 and 450 ppb for CH_4 (see Chapters 9 and 11).

Similar to the insolation trend in Figure 15-1, these proposed natural levels lie well toward the low extreme of their natural ranges and would be more favorable to a new glaciation than the observed pre-industrial

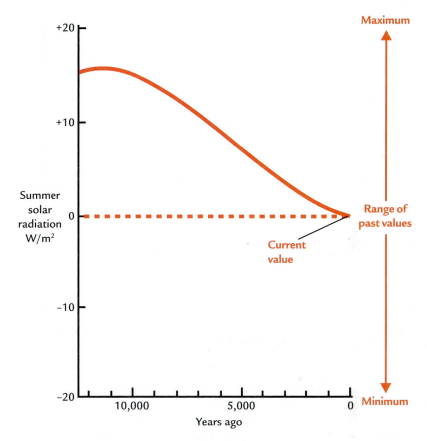

FIGURE 15-1 Summer insolation trend at 65°N during the last 12,500 years relative to the range of variation over the last several hundred thousand years. *[Insolation trend from A. Berger, "Long-term Variations of Caloric Insolation Resulting from Earth's Orbital Elements,"* Quaternary Research 9 (1978): 139–167.]

values. The hypothesized CO_2 drop to 240 to 245 ppm lies about 40% of the way toward the typical glacial-minimum concentration of 185 ppm over the last several hundred thousand years (Figure 15-2), while the proposed CH_4 drop to 450 ppb is about 70% of the way toward the low extreme of its natural range (not shown in the figure).

According to the early anthropogenic hypothesis, if the greenhouse-gas levels had followed their natural downward trends, Earth would have cooled enough by now to reach the threshold at which new ice sheets would have begun to form (Figure 15-3). Initially, this cooling would not have resulted in anything like a full ice age, within which ice sheets would have covered most of Canada and Scandinavia, because massive full-sized

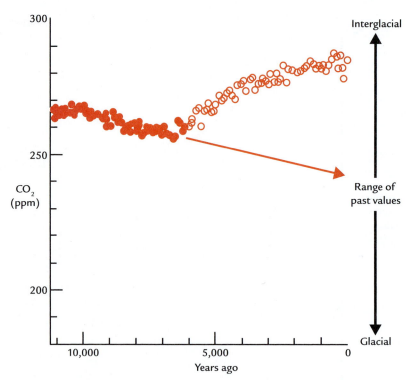

FIGURE 15-2 Observed *(circles)* and predicted *(red arrow)* CO_2 trend during the last 11,000 years relative to the range of variation over the last several hundred thousand years. *Open circles* mark anomalous CO_2 increase. *[CO$_2$ data from EPICA Community Members, "Eight Glacial Cycles from an Antarctic Ice Core,"* Nature *429 (2004): 623–628.]*

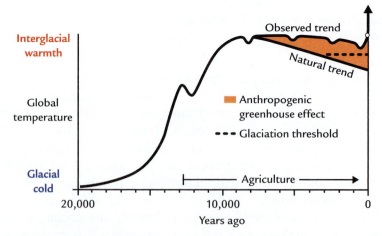

FIGURE 15-3 Hypothesized effect of pre-industrial greenhouse-gas emissions in keeping global temperature warm enough to avoid a glaciation that would have begun under natural conditions.

ice sheets take tens of thousands of years to form. Instead, this would have been the early phase of an ice age, with small ice sheets growing at higher and colder latitudes (and altitudes). But according to the anthropogenic hypothesis, greenhouse-gas emissions from pre-industrial farming stopped this process and averted a new glaciation.

Model Simulations

This part of the early anthropogenic hypothesis has been tested using quantitative models that simulate Earth's climatic responses (such as changes in air and ocean temperature, extent of snow cover, and amount of sea ice) to various factors that drive climate change. These model experiments can be thought of as "what-if" experiments: *What* would happen to Earth's climate *if* one or more of the driving factors in the model simulation were changed? In the case explored here, the altered factors are the lower CO_2 and CH_4 values proposed in the anthropogenic hypothesis. These reduced concentrations are entered into the model to determine how much they would have cooled Earth's climate, and whether or not new ice sheets would have started to form.

One kind of climate model, called an **EMIC**—short for "Earth system model of intermediate complexity"—can be used to simulate responses over long enough intervals of time (thousands of years) to evaluate whether or not ice sheets would have begun to grow. In order to run simulations over very long intervals of time (tens of thousands of years), these models make sacrifices in the way Earth is represented, such as using highly simplified, blocky-looking representations of continent/ ocean boundaries. These simplifications may reduce the accuracy of the simulations. Simulations run with EMIC models to date have given mixed results on the question of glacial inception: some show new ice beginning to form at the reduced CO_2 and CH_4 concentrations specified in the early anthropogenic hypothesis, but others do not.

Another EMIC simplification is of particular importance in testing the idea of glacial inception: mountains and plateaus are smoothed down to much lower elevations. Because of this smoothing, temperatures over the lower model terrain are warmer than those in real-world high terrain and are less covered by snow and ice. As a result, the failure to form new ice sheets in these simulations could be caused either by the failure of the hypothesis or by the limitations of the models.

A second type of model, called a **general circulation model**, represents the climate system more fully. Continents and higher-elevation areas are much more realistic-looking, although mountains and plateaus are still not as high as in the real world. Unfortunately, because of the large amount of computing power needed for these improved representations, this type of model is expensive to run and can only simulate a few decades, or at most a few centuries, of climate evolution. Because of this restriction, general circulation models cannot reproduce the slow growth and melting of ice sheets that occur over thousands to tens of thousands of years.

Instead, general circulation models can only simulate the number of months of snow cover present in each region over intervals of a few decades. Yet this information still has a direct bearing on the issue of glacial inception. Areas that show year-round snow cover that gradually thickens during the decades-long simulations are clearly headed toward even greater thicknesses of snow that will eventually consolidate into ice. At some point, the ice will thicken enough to begin to flow as glaciers. Without actually simulating ice sheets, the models reveal areas of year-round snow cover where ice sheets are likely to form.

Using General Circulation Models

The atmospheric scientists Steve Vavrus, John Kutzbach, and colleagues have run the largest number of experiments with general circulation models in order to test the early anthropogenic hypothesis. Their simulations have explored the kind of climate that would exist today at the lower CO_2 and CH_4 values predicted by the hypothesis. The central focus of this work has been whether the simulated cooling would permit regions of year-round snow cover to exist.

The experiments have used two generations of the community climate science model (CCSM) of the National Center for Atmospheric Research in Boulder, Colorado: an earlier version called CCSM3 and a more recent one called CCSM4. The experiments with CCSM3 were run with a CO_2 value of 240 ppm and a CH_4 value of 450 ppb, as proposed in the original hypothesis. The more recent experiments with the CCSM4 model used slightly adjusted values of 245 ppm for CO_2 and 445 ppb for CH_4. Both model simulations show a colder late pre-industrial world, with the CCSM4 cooling slightly less than that in CCSM3 because of the slightly higher CO_2 value used as an input to the simulation.

All climate model simulations show larger climatic responses at high latitudes than in the tropics because of the positive feedback effects from sea ice and snow (Figure 15-4). Increases in snow cover on land brighten the surface, compared to the darker brown and green hues of snow-free land. Over the ocean, advances in white or gray-white sea ice brighten the surface compared to the darker blue of the liquid surface (Figure 15-5). As these brighter surfaces grow in extent, they reflect a greater fraction of the incoming solar radiation rather than absorbing

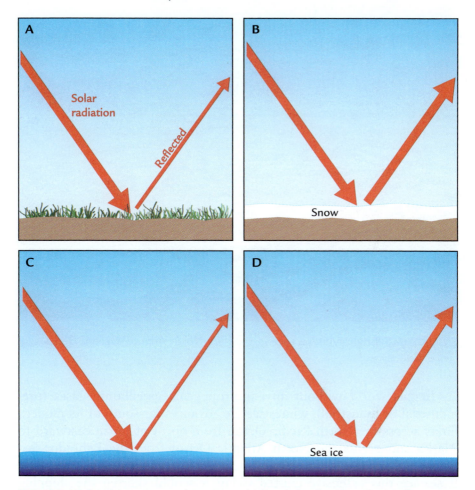

FIGURE 15-4 White snow and sea ice at higher latitudes (B and D) reflect more solar radiation, cooling the lower atmosphere, compared to snow-free land and ocean (A and C).

it, and the resulting loss of solar heating amplifies the cooling of the surrounding region. This process is called **albedo-temperature feedback**.

Because the extent of snow and sea ice is largest in winter, positive feedback amplifies the size of the global-average changes far more in that season than in summer, when snow and sea ice cover is smaller. In the experiments examined here, the reductions in CO_2 and CH_4 levels resulted in high-latitude winter climates that were colder by an average of 5°C (9°F) or more compared to the modern era. This difference in winter temperature results from early greenhouse-gas emissions due to farming and emissions during industrial time.

FIGURE 15-5 Bright-white sea ice in the dark North Atlantic Ocean. *[Sylvain Cordier/Getty Images.]*

In this world without any anthropogenic greenhouse gases (pre-industrial or industrial), winter sea ice advances far south of its modern limits across the North Pacific Ocean from north of Japan eastward to Alaska and across the North Atlantic Ocean from the Labrador Sea (between Canada and Greenland) eastward to Scandinavia (Figure 15-6). If sea ice this extensive existed today, it would shorten the shipping season in these ocean areas, as well as in inland seas such as Hudson Bay in central Canada and the Baltic Sea between Scandinavia and mainland Europe. Sea ice advances tend to be larger in the North Pacific Ocean in the CCSM3 simulation, but larger in the northwest Atlantic Ocean in the CCSM4 simulation. These differences are typical of the normal variability among different model simulations.

The summer cooling simulated by the two experiments is smaller than that in winter because snow and sea ice are much less extensive, and their albedo feedback effect is reduced. On average, summer temperatures over the higher latitudes of the northern continents would have been 2° to 2.5°C (3.5° to 4.5°F) cooler than in the modern era. A cooling of this size would have reduced the length of the frost-free growing season by several weeks in both spring and autumn, making agriculture

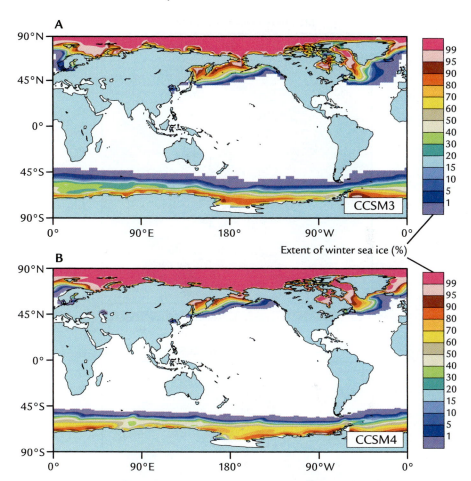

FIGURE 15-6 Simulated extent of winter sea ice for two model experiments (CCSM3 and CCSM4) using the lower CO_2 and CH_4 values proposed in the early anthropogenic hypothesis. *[Simulations are from in-progress work by S. Vavrus, F. He, and J. Kutzbach.]*

difficult or impossible in some far-northern areas of the North American plains (in Canada and the United States), western (European) Russia, and eastern Europe.

As summarized in Chapter 1, summer is the critical season that determines whether the previous season's snow survives and thickens into glacial ice during subsequent years. Both the CCSM3 and CCSM4 simulations show substantial areas in which snow cover persists through all twelve months of the year: northwest North America and northeast

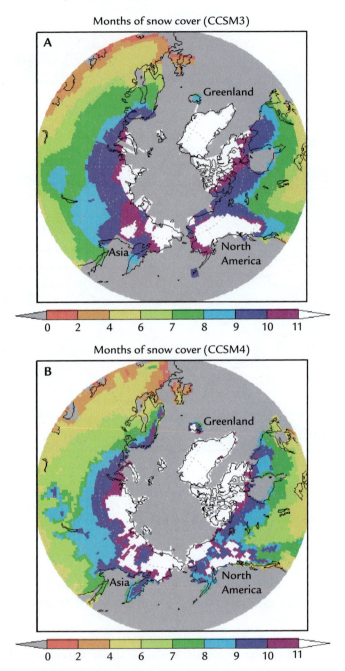

FIGURE 15-7 Simulated months of summer snow cover for two model experiments (CCSM3 and CCSM4) using the lower CO_2 and CH_4 values proposed in the early anthropogenic hypothesis. *[Simulations are from in-progress work by S. Vavrus, F. He, and J. Kutzbach.]*

Canada, the Arctic islands north of mainland Eurasia, and eastern Siberia (Figure 15-7). The high-Arctic region in far northeastern Canada, which has year-round snow cover in both model simulations, has long been thought to be an initial "breeding ground" for the later growth of the huge North American ice sheet.

In both simulations, the total area in the Northern Hemisphere with year-round snow cover is larger than the present-day area of the Greenland ice sheet. Because permanently snow-covered regions are considered to be areas of incipient glaciation, both simulations provide support for the claim in the early anthropogenic hypothesis that ice sheets would be forming and advancing today, had it not been for anthropogenic greenhouse-gas effects.

But what kind of "ice sheets" would these be? As snowfields pile up to depths of tens of meters, the compressed snow gradually turns to ice. Glacier experts have found that, under favorable conditions, the annual addition of snow each year can be as great as half a meter (1.5 feet) of solid ice. Once the ice reaches a thickness of 50 meters after a century or more, it would become capable of flowing downhill under its own weight as a glacier. If the ice accumulation continued for 1,000 to 2,000 years, the ice sheets growing in these areas might reach a thickness of as much as 500 to 1,000 meters (about 1,500 to 3,000 feet). By way of comparison, the modern Greenland ice sheet is over 2 kilometers thick, and the ice sheet on North America 20,000 years ago was over 4 kilometers thick. So this world of smaller and thinner ice sheets would be far short of a fully glacial one, but it would no longer be a fully interglacial one either.

Future work will refine these models and likely alter their simulations of permanent snow cover and glacial inception. One factor that will be improved in the future is model resolution, as measured by the size of the individual **grid boxes** within which the models calculate average climatic responses. The results from the CCSM3 model are based on an average grid-box size of 2.8 degrees latitude by 2.8 degrees longitude, while those from the CCSM4 model are based on a grid-box size of 1.4 by 1.4 degrees (roughly 160 kilometers by 160 kilometers, or 100 miles by 100 miles). The higher-resolution CCSM4 model reproduces more of the high topography found in areas like the northern Rockies and the Coast Range and Brooks Range in Alaska. As a result, the year-round snow cover simulated by CCSM4 is more obviously confined to these areas of higher topography than it was in the lower-resolution CCSM3 version (see Figure 15-7). But even the 100-by-100-mile resolution still smoothes most of the narrow mountain peaks (Figure 15-8).

FIGURE 15-8 Mountain glaciers and snow cover in high Rocky Mountain terrain. *[David Nunuk/Photo Researchers.]*

As improvements in computing power continue to reduce the size of grid boxes, general circulation models will do a better job of resolving high topography and areas of likely snow cover under the lower greenhouse-gas concentrations specified in the early anthropogenic hypothesis. Climatic processes that are still being added to or improved in models (such as changes in ocean circulation and sea ice, as well as different kinds of vegetation on the continents) will also affect future simulations. Estimates of "natural" CO_2 and CH_4 levels during pre-industrial times may also be adjusted as new evidence emerges.

Summary

Despite the above caveats, the evidence from climate model simulations at this point tentatively favors a key prediction of the early anthropogenic hypothesis. If CO_2 and CH_4 concentrations had fallen to the levels proposed in the hypothesis, the early stages of a new glaciation would be underway. Greenhouse-gas emissions from early farming appear to have prevented the onset of a new glaciation.

Additional Resources

Claussen, M., V. Brovkin, R. Calov, A. Ganapolski, and C. Kubatzki. "Did Humankind Prevent a Holocene Glaciation?" *Climatic Change* 69 (2005): 409–417.

Kutzbach, J. E., W. F. Ruddiman, S. J. Vavrus, and G. Philippon. "Climate Model Simulation of Anthropogenic Influence on Greenhouse-induced Climate Change (Early Agriculture to Modern): The Role of Ocean Feedbacks." *Climatic Change* 99 (2010): 351–381. doi:10.1007/s10584-009-9684-1.

Vavrus, S. J., G. Philippon-Berthier, J. E. Kutzbach, and W. F. Ruddiman. "The Role of GCM Resolution in Simulating Glacial Inception." *Holocene* 21 (2011): 819–830. doi:10.1177/0959683610394882.

Vettoretti, G., and W. R. Peltier. "The Impact of Greenhouse Gas Forcing and Ocean Circulation Changes on Glacial Inception." *Holocene* 21 (2011): 803–817. doi:10.1177/0959683610394885.

OTHER CLIMATIC EFFECTS OF EARLY LAND CLEARANCE

E vidence summarized in Chapter 8 showed that early clearance of forests in Eurasia and the Americas was much more extensive than scientists once thought. The forest vegetation that was burned during the slow clearing process across these regions contributed large amounts of CO_2 to the atmosphere during pre-industrial times and made pre-industrial climate warmer than it would otherwise have been, but CO_2 (and CH_4) emissions were not the only climatic consequences of early clearance. In addition, the transformation of forests to croplands and pastures altered the way in which large areas of Earth's surface responded to incoming solar radiation, thereby changing regional climate. Removing the forest cover also potentially injected more dust into the atmosphere, altering climate even further.

Effects of Land Clearance on Incoming Solar Radiation

Large areas of Earth's continents would be covered by forest under natural conditions (Figure 16-1), except for tundra at high northern latitudes and grasslands and deserts at lower latitudes. When these darker forests are replaced by lighter pastures or croplands, more incoming solar radiation is reflected from the newly brightened surfaces, rather than being absorbed. The climatic effects from these kinds of

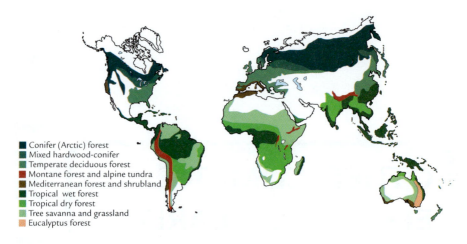

Conifer (Arctic) forest
Mixed hardwood-conifer
Temperate deciduous forest
Montane forest and alpine tundra
Mediterranean forest and shrubland
Tropical wet forest
Tropical dry forest
Tree savanna and grassland
Eucalyptus forest

FIGURE 16-1 Natural global forest vegetation. *[Adapted from R. L. Smith and T. M. Smith, Elements of Ecology (Menlo Park: Benjamin/Cummings, 1998).]*

changes vary with latitude because of differences in topography and the amount of snow cover.

Large land areas above 50°N in North America and 55°N in Eurasia are covered by dark evergreen conifer forests of spruce and fir. These trees are "evergreen" in that they retain their needles through the year, even as new ones are pushing out and replacing old ones. Seen from above, these forests are a dark green color year-round, including in winter.

At Arctic and near-Arctic latitudes, snowfall is plentiful in winter and through much of the spring and autumn. When snow falls on Arctic forests, strong winds dislodge it and dump it on the ground beneath, leaving most of the canopy dark green. In comparison, the Arctic tundra vegetation farther north consists of low-lying shrubs and grasses. When snow falls on this tundra, it blankets the vegetation and turns the surface bright white.

Natural changes from darker forest cover to brighter tundra increase the amount of sunlight reflected back into space (Figure 16-2). During interglacial times like today, when the climate is warm, Arctic forests extend well to the north, and the dark green canopy masks the cold-season snow cover lying underneath. As a result, the Sun's radiation absorbed by the forest canopy warms the air. But when climate cools, the dark evergreen forests retreat south, giving way to brighter tundra. Solar radiation that had previously been absorbed by the dark forest canopy is now reflected by the light tundra, and the local climate cools even more during the long snowy season. This enhancement of

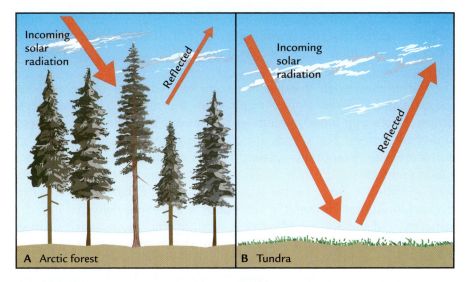

FIGURE 16-2 The dark canopies of evergreen Arctic forest absorb solar radiation (A). Replacement of forests by snow-covered pasture enhances reflection of solar radiation and causes cooling (B).

natural cooling by changes in vegetation is called **vegetation-albedo feedback**. The same cooling process occurs when humans clear high-latitude conifers to make room for crops or pastures.

At middle northern latitudes (between 25° and 55°), hardwood forests of oak, hickory, maple, ash, elm, beech, and other species thrive in well-watered regions (see Figure 16-1). These trees lose their leaves in autumn just prior to the first snows and put out new leaves in spring after the last major snowfalls. Because they have no leaves during the snowy season, deciduous trees don't mask the underlying ground from solar radiation as much as Arctic forests do; although, small amounts of evergreen trees (pine, cedar, and hemlock) in some mid-latitude regions have some masking effect. In addition, snow cover becomes progressively less frequent toward lower mid-latitudes. As a result, even though deciduous forests absorb more incoming radiation in winter than grasslands or croplands, the effect on regional temperature is not as large as at high northern latitudes. Clearing of deciduous forests for agriculture causes a small cooling in winter, but little change in summer.

At latitudes between 25°N and 25°S, tropical evergreen forests (rain forests) dominate the natural vegetation in areas of plentiful rainfall (see Figure 16-1). Because of the absence of hard freezes, tropical wet forests remain green year-round and retain leaf cover even while shedding old leaves. Because snow does not fall in the tropics except in small areas of high-mountain terrain, its effect on the reflectivity of tropical vegetation is negligible.

The differences in response between forested lands and grasslands (or croplands) in the tropics result from several opposing processes. Tropical trees pull large amounts of soil moisture up through their root and trunk systems and recycle it into the atmosphere as water vapor in a process called **evapotranspiration**. The water vapor sent into the atmosphere condenses and forms clouds with bright upper surfaces that reflect solar radiation and keep some of it from reaching Earth's surface. When tropical forests are cleared, the amount of water vapor delivered into the atmosphere decreases, cloud cover is reduced, and more solar radiation reaches the ground, which tends to warm the climate (Figure 16-3).

But clearing tropical forests also converts darker tree canopies to brighter pasture and crop surfaces (see Figure 16-3). The resulting increase in reflection of incoming solar radiation from surfaces cleared for agriculture cancels part of the increase in incoming radiation. Ultimately, because croplands and grasslands are darker than the very bright cloud tops, land clearance in the tropics results in a small increase in solar radiation absorbed, and a small warming of climate.

In summary, clearance of Arctic forests for agriculture makes high latitudes cooler during the long winter season, mid-latitudes slightly cooler during the short winter season, and the tropics slightly warmer, mainly in the moist summer season.

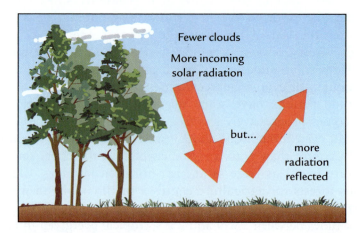

FIGURE 16-3 Clearance of tropical trees reduces cloud cover and lets in more solar radiation, but the cleared ground reflects more of the radiation.

Predicted Climatic Effects of Pre-Industrial Forest Clearance

The climatic effects of pre-industrial forest clearance depend mainly on two factors: (1) the regions (latitudes) in which pre-industrial clearance occurred, and (2) the natural vegetation in those regions prior to clearance. The estimated clearance for 2,000 years ago in Figure 16-4 was determined by Jed Kaplan and colleagues, and the natural distribution of forest vegetation can be seen in Figure 16-1.

Long before the pre-industrial era, by around 40,000 years ago, people knew how to sew tight-fitting warm clothing and build dwellings made of stone, bone, or wood to help them survive at higher, colder latitudes. Yet almost no one lived in or cleared high-latitude forests through most of pre-industrial time because the climate was so cold that the growing season was very short. As a result, early clearance north of 55°N in Eurasia was extremely limited, except along the southernmost conifer forest fringe in western (European) Russia after 1400. The spruce forests in present-day Canada and the northernmost United States also remained largely intact through late pre-industrial times (until 1700) because only a few million people lived in North America, mainly in the Mississippi River valley well south of the Arctic forests (see Chapter 6, Figure 6-16).

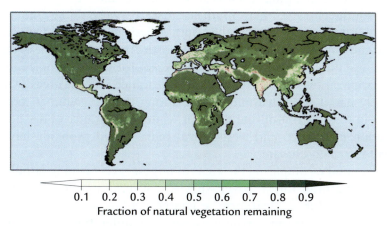

| 0.1 | 0.2 | 0.3 | 0.4 | 0.5 | 0.6 | 0.7 | 0.8 | 0.9 |

Fraction of natural vegetation remaining

FIGURE 16-4 Simulated clearance of natural vegetation 2,000 years ago. *[Adapted from J. O. Kaplan, K. M. Krumhardt, E. C. Ellis, W. F. Ruddiman, C. Lemmen, and K. K. Goldewijk, "Holocene Carbon Emissions as a Result of Anthropogenic Land Cover Change," The Holocene 25 (2011), 775–791, doi:10.1177/0959683610386983.]*

Because clearing of high-latitude forests was negligible for most of the last 7,000 years, the net amount of cooling during the long snow season should have been minimal. After the year 1500, but before the Industrial Era, cutting of forests in western Russia and the eastern Baltic Sea area accelerated as western European nations that had by then been fully deforested turned to these more remote regions for lumber (see Chapter 4, Figure 4-18). This forest clearance likely cooled winter temperatures in the last few centuries of the pre-industrial era, but had little effect on summer temperatures.

During pre-industrial times, more people lived at mid-latitudes (25° to 55°), where climates were temperate and the frost-free growing season was longer. Several mid-latitude regions were very heavily populated: northern China, Mesopotamia (the eastern Mediterranean), and Europe. Snow falls in high northern mid-latitudes in winter, but melts away for well over half of the year, while little snow falls at lower mid-latitudes. The natural forests in these mid-latitude regions are mostly deciduous trees which, except for local stands of pine, cedar, and hemlock, have a small snow-masking effect. Overall, early clearance of mid-latitude forests in China, Europe, and Mesopotamia likely had a small cooling effect, mainly in winter. Extensive clearance of the deciduous forests in North America should only have had climatic effects after 1700.

Most pre-industrial people lived in the tropics (between latitudes 25°N and 25°S), because the warm frost-free climate permitted two or even three crop plantings each year. Heavily populated regions included India, south-central China, Southeast Asia between India and China, Mexico and Central America, the South American lowlands between the Andes and the Amazon, and tropical North Africa. All of these regions were covered by moist tropical forests except for the dry tropical forests in west-central India and Mexico. Early clearance in these areas by farmers likely had a small warming effect during the wet summer monsoon season, but little effect during the drier winters. Clearance of dry-adapted vegetation in Mexico and west-central India would probably have had a small effect on climate.

In summary, comparing the areas of largest early farming populations with the natural forest cover suggests several predicted effects of pre-industrial forest clearance: a winter cooling in the very limited regions at high latitudes where evergreen conifer forests were cut; a smaller winter cooling across large areas in Europe, Mesopotamia, and northern China where deciduous forests were removed; and a small warming across areas of the tropics where forests were cleared.

Model Simulations

The climate modeler Julia Pongratz and colleagues were the first scientists to simulate the climatic effect of pre-industrial land clearance. Their experiments found a very small net global cooling of about 0.05°C because of increased reflection of solar radiation at higher latitudes. But their simulation was based on the assumption of very small pre-industrial clearance (see Chapter 8), and it may have underestimated the size of the climatic effect.

The climate modeler Feng He and colleagues later ran experiments based on the larger amount of agricultural clearance for the pre-industrial era simulated by Jed Kaplan and colleagues (recall Chapter 8). They compared the climatic effect of this clearance to the baseline of natural (forest) vegetation in the absence of humans.

The net annual global temperature change of –0.17°C they simulated for the end of the pre-industrial era (1850) was larger than the estimate from Pongratz and colleagues, but still just a relatively small fraction (12% to 13%) of the 1.3°C to 1.4°C warming estimated from model simulations of the proposed agricultural increases in pre-industrial greenhouse gases in the early anthropogenic hypothesis.

In some high-latitude areas, the simulated cooling caused by albedo changes by the year 1850 reached values of 1°C or more, large enough to cancel as much as 20% of the pre-industrial greenhouse-gas warming. But much of this high-latitude albedo cooling occurred very late in pre-industrial time because of deforestation in northern Eurasia after 1400 (see Chapter 4), and in east-central North America after 1700 (see Chapter 6).

For a longer-term perspective on the albedo effect of early agricultural clearance, Feng He and colleagues also ran a simulation for 2,000 years ago (Figure 16-5). By that time, the rise in atmospheric CO_2 concentration that began 7,000 years ago had already reached, or was very close to, its 22-ppm peak. As a result, most of the early anthropogenic temperature increase was already in place, including the warming of 2° to 5°C across northern mid-latitudes and higher latitudes.

Feng and colleagues found that clearance by the year 2000 at middle and high latitudes caused an estimated albedo-related annual cooling that generally varied between 0° and 0.5°C. This estimate is a relatively small fraction (about 10%) of the 2° to 5°C annual warming simulated for the greenhouse-gas increases at those latitudes.

In the regions at very high latitudes where the model experiments described in Chapter 15 simulated the persistence of year-round snow cover and incipient glaciation because of reduced greenhouse gases

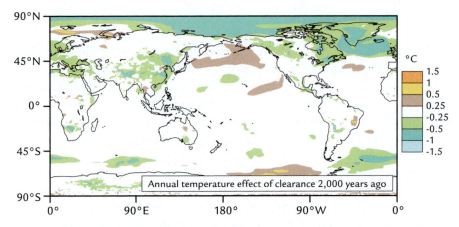

FIGURE 16-5 Simulated annual mean temperature changes at Earth's surface caused by forest clearance 2,000 years ago. *[Simulations are from in-progress work by F. He, S. Vavrus, and J. Kutzbach.]*

(see Chapter 15, Figure 15-7), it seems unlikely that this small opposing albedo effect would have offset much of the greenhouse-gas effect. Future modeling experiments that combine both influences are needed to determine what fraction of the permanent snow cover simulated in Figure 15-7 persists.

Effects of Clearance on Atmospheric Dust (Aerosols)

Early farming and related human activities could also have had an effect on climate by sending several kinds of aerosol particles into the **troposphere**, the lower 10 to 20 kilometers (6 to 12 miles) of Earth's atmosphere, depending on seasonal changes. Forest clearance destabilizes the landscape by exposing soils to winds that send dust particles into the air. Dust particles are often carried downwind in the prevailing circulation across regions where they have a small cooling effect by blocking incoming solar radiation, but most of the particles are washed out of the air by rain before they travel far from their sources.

During the Industrial Era, aerosols emitted by land clearance have caused a cooling that has counteracted part of the greenhouse-gas warming. Mechanized farming has been responsible for increased dust from soils because of deep plowing in semi-arid prairie and steppe

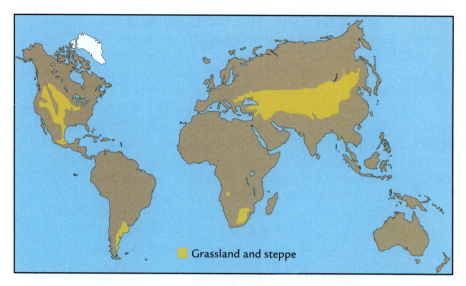

FIGURE 16-6 Distribution of semi-arid prairie and steppe regions converted to agriculture since the advent of mechanized farming, around 1850. *[Adapted from R. L. Smith and T. M. Smith,* Elements of Ecology *(Menlo Park: Benjamin/Cummings, 1998).]*

regions (Figure 16-6). Prior to mechanized agriculture, these regions were covered by deep-rooted prairie vegetation (herbs and grasses) that were adapted to survive even severe droughts, but their tenacious roots also prevented the cultivation of crops. Mechanized plows capable of tearing up the deep roots brought farming to the prairies and steppes during the late 1800s, but also exposed the land surface to strong winds during dry intervals. The Dust Bowl of the 1930s in the semi-arid American Great Plains occurred when a severe drought struck areas where prairie soils had been disturbed by plowing.

In contrast, dust produced by plowing in pre-industrial time is likely to have played only a minor climatic role. Prior to mechanized agriculture, farming was mostly restricted to naturally wet, forested regions in Europe, Southern Asia, the Americas, and Africa. Forest clearance in naturally wet regions was less likely to expose soils to extreme drying and lead to injection of aerosols into the lower atmosphere. Sediment cores from lakes and rivers show increased deposition of eroded silt and clay in recent millennia (see Chapter 4, Figure 4-11), but most of this debris was likely moved by water, not wind. Early farming in semi-arid regions was largely restricted to those areas supplied by reliable sources of irrigated river water derived from snowmelt or rainfall in nearby mountains. Areas with less reliable supplies of water would have been more vulnerable to extreme droughts, but these regions were not extensively farmed.

A second kind of aerosol emitted today comes from cook stoves used by people in areas of southern and eastern Asia. These devices burn any combustible material available (often cow dung), and they emit regional-scale plumes of carbon-rich smoke into the lower atmosphere. These **brown clouds** trap large amounts of incoming solar radiation a kilometer or so above the land surface and warm that layer of atmosphere, but they block enough solar radiation to cool the land surface below.

Emissions of this kind of aerosol during pre-industrial time are likely to have been closely linked to population. By the year 1000, the population of Asia was still less than 5% of the present level, and it remained below 10% of current levels until after the year 1700. As a result, brown-cloud aerosol particles likely had relatively little regional-scale effect on global climate during most of pre-industrial time.

In contrast to the first two kinds of aerosol particles, a third kind of emissions could have played a more significant pre-industrial climatic role—aerosols derived from burning of biomass to clear debris from cultivated areas and pastures (see Chapter 8). Evidence from modern geographic patterns suggests that biomass burning on a per capita basis was much larger when population densities were smaller, and then would have decreased as populations grew. Quantifying the climatic role of past aerosols from biomass burning is difficult; even their modern-day effects remain the largest uncertainty in estimating current human effects on climate. Part of the problem is that these aerosols interact with the climate system in many different ways. The effect of these aerosols on pre-industrial climate needs further investigation.

Some of the soot from both high-latitude biomass burning and from natural fires set by lightning falls on Arctic sea ice and makes its surface darker. The resulting increase in absorption of solar radiation helps to melt the ice. During most of pre-industrial time, however, the minimal human presence at very high latitudes likely held the effect of anthropogenic soot on Arctic sea ice to a minimum.

The major effect of biomass burning at middle and lower latitudes would have likely been a net cooling, partly because the particles block incoming solar radiation and partly because they form tiny centers around which atmospheric water vapor can condense and form low-elevation clouds that also block solar radiation. In pre-industrial times, this cooling effect would likely have been limited to regions around, and downwind from, the burning sites at middle and lower latitudes. Whatever their pre-industrial effect on lower latitudes, it seems unlikely that biomass burning would have had much effect on the high-latitude regions where model simulations show enough cooling to permit permanent snow cover to form (see Chapter 15, Figure 15-7).

Other aerosols that cool climate in the modern world, such as **sulfate aerosol** particles from burning coal, were much less prevalent during the pre-industrial era. Early increases in emissions from China during the last 1,500 years, and in a few countries of northern Europe during the last several centuries just prior to the Industrial Era, were small in size compared to modern levels.

Summary

Model simulations based on land-use reconstructions indicate that early agricultural clearance caused a small albedo-related cooling in the Northern Hemisphere at middle and high latitudes. Early agricultural emissions of aerosols may have added to this cooling. But these two cooling effects are unlikely to have offset more than a small fraction of the warming caused by rising agricultural emissions of greenhouse gases. In view of the much larger greenhouse-gas climatic warming effects at very high latitudes, it appears likely that year-round snow cover (glacial inception) will persist at some level in future model simulations that assess the combined effects of greenhouse gases, forest clearance, and aerosols. The results of simulations to date still support the conclusion reached in Chapter 15 that early agricultural greenhouse gases prevented the start of a new cycle of glaciation within the last few millennia.

Additional Resources

Bala, G., K. Caldeira, M. Wickett, T. J. Phillips, D. B. Lobell, C. Delire, and A. Mirin. "Combined Climate and Carbon-cycle Effects of Large-scale Deforestation." *Proceedings of the National Academy of Sciences* 104 (2007): 6550–6555.

Bonan, G. B. "Forests and Climate Change: Forcings, Feedbacks, and the Climate Benefits of Forests." *Science* 320 (2008): 1444–1449. doi:10.1126/science.1155121.

Pongratz, J., C. H. Rieck, T. Raddatz, and M. Claussen. "Biogeophysical Versus Biogeochemical Climate Response to Historical Anthropogenic Land Cover Change." *Geophysical Research Letters* 37 (2010), LO8702. doi:10.1029/2010GL043010.

CHAPTER 17

THE END OF NORTHERN HEMISPHERE GLACIATIONS

E arth's climate changes at many time scales, with shorter-term changes superimposed as wiggles on top of other longer-term oscillations, just as day-to-day temperature fluctuations ride atop longer seasonal changes today. Over the longest time scales (tens of millions of years), evidence from a range of sources shows that Earth has been taking a very gradual trip into the refrigerator (Figure 17-1). Following a time of much greater warmth 50 million years ago, the planet has cooled by a large amount, with ice sheets appearing first in southern and then in northern polar regions. Several factors may have contributed to this cooling, but most scientists infer that slowly falling CO_2 levels were the primary cause.

The pole-centered continent of Antarctica reached a refrigerated state long before the northern continents located at lower, more temperate latitudes. During the relative warmth of 50 million years ago, beech trees were common on Antarctica (even though it was in its modern-day polar location), but by 35 million years ago climate had cooled enough for the first ice to appear. By 14 million years ago, Antarctica had entered a deep freeze, with ice permanently present on much of the continent. Now, even during times of interglacial warmth like today, ice extends all the way to the coasts along most of this polar continent.

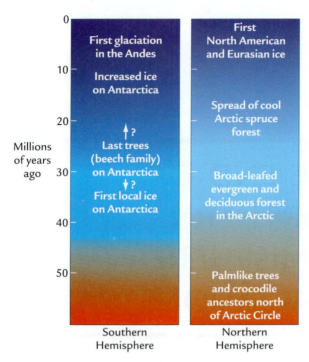

0

First glaciation
in the Andes

First
North American
and Eurasian ice

10

Increased ice
on Antarctica

Spread of cool
Arctic spruce
forest

20

↑ ?

Last trees
(beech family)
on Antarctica

↓ ?

First local ice
on Antarctica

Broad-leafed
evergreen and
deciduous forest
in the Arctic

Millions
of years
ago 30

40

50

Palmlike trees
and crocodile
ancestors north
of Arctic Circle

Southern
Hemisphere

Northern
Hemisphere

FIGURE 17-1 Earth's gradual cooling trend over the last 50 million years is evident from records of vegetation types and ice-sheet development in both polar regions.

A simultaneous cooling trend occurred in the Arctic, starting with a warm Arctic Ocean with no sea ice 50 million years ago and ending with today's ice-covered ocean. On the northern Arctic margins of Eurasia and North America, stands of warm tropical and semi-tropical vegetation 50 million years ago gradually gave way to cooler deciduous forests, then to cold Arctic forests of spruce and larch, and finally, within the last few million years, to tundra.

Small mountain glaciers began to form on Greenland around 7–8 million years ago, and ice sheets of substantial size first appeared on North America and Scandinavia near 2.75 million years ago. The subsequent ice-sheet advances and retreats at orbital cycles were superimposed over a long, slow cooling trend (Figure 17-2). From 2.75 to 0.9 million years ago, moderately large ice sheets grew and melted mainly at the 41,000-year rhythm of orbital tilt (recall Chapter 1), with ice present roughly half of the time and absent the other half. During the last 1 million years, ice sheets have varied mainly at a 100,000-year rhythm,

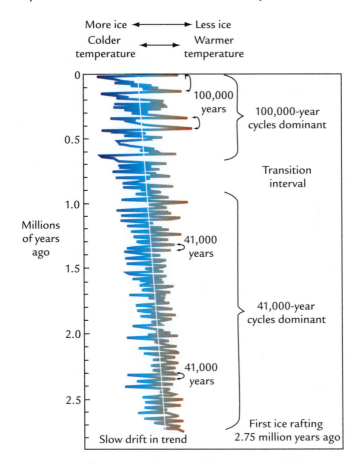

More ice ⟵——→ Less ice
Colder ⟷ Warmer
temperature   temperature

FIGURE 17-2 Past oscillations in the size of northern hemisphere ice sheets at cycles of 100,000, 41,000, and 22,000 years. *[Adapted from M. E. Raymo, "The Initiation of Northern Hemisphere Glaciation," Annual Reviews of Earth and Planetary Sciences 22 (1994): 353–383.]*

have reached larger sizes, and have been present on North America for more than 90% of the time. This progression from no ice sheets, to ice present about half the time, to ice present more than 90% of the time, marks the slow northern hemisphere cooling (see Figures 17-1 and 17-2).

The Next Glaciation?

What about the next glaciation? Ignoring for now the large greenhouse-gas overprint underway during the current Industrial Era, when would the next glaciation arrive under natural conditions? A popular view described at the start of Chapter 15 holds that the start of the next natural glaciation could occur within the next few thousand years, as summer insolation levels fall slightly beyond the already low present-day values.

Other scientists have drawn the very different conclusion that Earth's natural glacial cycles in the north have "skipped a beat" (Figure 17-3). Summer insolation in the Northern Hemisphere is now at, or very close to, a short-term minimum value that is as favorable to initiating a new glaciation as any level that will occur during the next 50,000 years. Yet no new glaciation has begun. With both North America and Eurasia ice-free for the last 7,000 years and perhaps for the next 50,000 years, this proposed "skipped beat" would eventually end up lasting 55,000 to 60,000 years.

But does this idea really make sense? No interglacial interval lasting that long (55,000–60,000 years) has occurred in the entire history of northern hemisphere glaciations covering 2.75 million years (see Figure 17-2). For roughly the first 2 million years of this interval, glaciations came and went every 41,000 years, with interglaciations lasting for about half of each cycle, or about 20,000 years. For most of the last 1 million years, no interglaciation has lasted longer than 10,000 years, not even the stage 11 interglaciation near 400,000 years ago (see Chapter 9, Figure 9-7). Given that no interglaciation in the

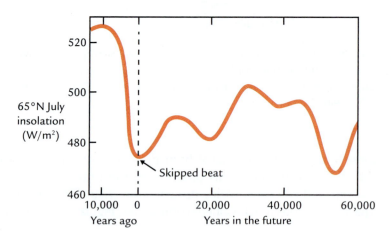

FIGURE 17-3 With present-day summer insolation levels in the Northern Hemisphere already close to the lowest values that will occur in the next 50,000 years, have the northern glacial cycles "skipped a beat"? *[Insolation values from A. Berger, "Long-term Variations of Caloric Insolation Resulting from Earth's Orbital Elements," Quaternary Research 9 (1978): 139–167.]*

50 or so glacial-interglacial cycles of the last 2.75 million years has lasted for anything close to 55,000 years, does it really make sense that Nature has just recently veered off into such unprecedented behavior?

The evidence in Chapter 15 provides a plausible alternative view: Nature would actually have caused the start of a new glaciation by now, but humans inadvertently intervened and stopped this from happening. If new ice had begun to form in the last 2,000 years or so, the stage 1 interglaciation would have lasted for less than 7,000 years, consistent with the short length of preceding interglaciations during the last 1 million years. In effect, the glaciation cycles have indeed "skipped a beat," but the deciding factor was human intervention, not natural causes.

The Greenhouse Era: No Future Glaciations

In the current Industrial Era, greenhouse-gas concentrations in the atmosphere are rising rapidly. As a result, Earth has begun to warm, and we will not enter the next glacial cycle for as long as our gas emissions continue to override the natural climate system. The warmer temperatures in the future will not just prevent the growth of new ice sheets, but will also melt parts of the existing ones. The Greenland ice sheet is vulnerable to melting because it lies at low enough latitudes to be in contact with relatively warm air masses and warm oceans. Although the West Antarctic ice sheet lies in a polar region where very cold air masses prevail, large parts of it float in an ocean that contains enough heat to melt ice. The much larger East Antarctic ice sheet is less vulnerable to melting because of its great height and its limited contact with the ocean.

Greenhouse gases will have a very large climatic impact on this planet in the coming centuries, although the size of this effect remains highly uncertain (Figure 17-4). Currently, the CO_2 concentration in the atmosphere is close to 400 ppm, but future levels are projected to build to a peak of at least twice (2x) the size of the 280-ppm pre-industrial value (500–600 ppm) and possibly as much as four times (4x) that level (1,100–1,200 ppm). The CO_2 concentration increases for this range of possible scenarios will warm global climate by almost 3°C to 6°C (5.5°F to 11°F) beyond the pre-industrial level (Figure 17-5). At high latitudes, the future warming will be at least twice the global average because of positive feedback from melting snow and sea ice, warming by an additional 6°C to 12°C (11°F to 22°F).

The high-end extreme of this projected range of future warming could occur if we humans fail to change our carbon-emitting habits, and

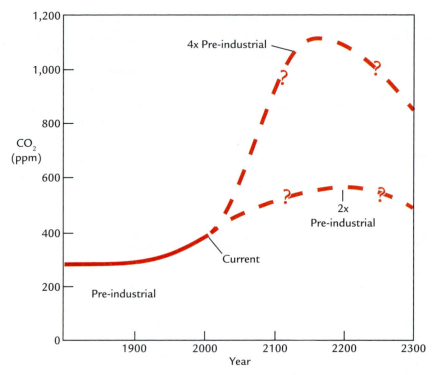

FIGURE 17-4 Changes in atmospheric CO_2 concentrations for two future scenarios: no action on CO_2 emissions (the 4x pre-industrial scenario), and major reductions in CO_2 emissions (the 2x pre-industrial scenario).

if new technologies do not come to our rescue. The lower end of the range could occur even if we manage to reduce our carbon emissions significantly, either through conservation or technological innovation. Many scientists think that this entire range of possible future CO_2 concentrations (500–1,200 ppm) poses serious risks for humans and for Earth's ecosystems. In both projections (the 2x pre-industrial and 4x pre-industrial scenarios, recall Figures 17-4 and 17-5), portions of the Greenland and West Antarctic ice sheets will be much more vulnerable to rapid melting than they are now.

The major effect of fossil fuels on future climate will be caused by CO_2 emissions. The accessible amount of oil still buried in the ground will begin to decline within a decade or two despite technological advances, but at least some oil will remain worth extracting through the end of the century. Natural gas reserves may last decades longer because of new extraction methods like **hydrofracking**, but they will not remain recoverable forever. The vastly larger amounts of carbon in coal and tar

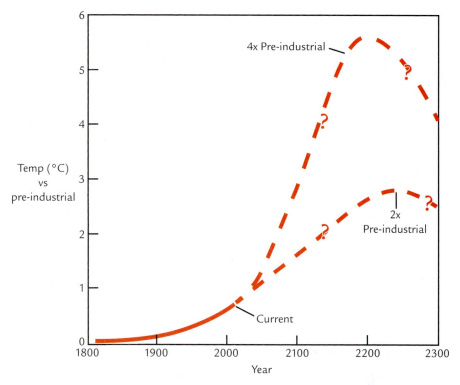

FIGURE 17-5 Changes in average global temperature for the 4x pre-industrial and 2x pre-industrial CO_2 scenarios.

sands will last for as much as two centuries before becoming uneconomical to extract. If we burn through all of these resources, we will drive atmospheric CO_2 concentrations toward or beyond the high end of the projected range (about 1,100–1,200 ppm).

Regardless of how much additional carbon dioxide is added, as the anthropogenic "pulse" of CO_2 we put in the atmosphere builds to peak levels in the not-too-distant future, our descendants will be living in a much warmer world than the one we live in now. After one or two centuries, the rate at which we burn carbon will slow, and the rate of absorption of our fossil-fuel CO_2 pulse in the ocean will outweigh our emissions. Atmospheric CO_2 concentrations will then fall, and Earth will begin to cool, but it will still remain much warmer than today.

Climate trends much farther in the future, thousands to tens of thousands of years from now, are hard to predict, mostly because of uncertainties in how future human activities will alter the natural climate system. One important question is whether or not CO_2 concentrations will drop to values low enough to cool Earth and allow ice sheets to return to the Northern Hemisphere.

Consider one possibility—that our carbon emissions drive atmospheric CO_2 concentrations to 800–900 ppm, or about mid-way between the extreme scenarios shown in Figure 17-4. This level would be 520 to 620 ppm above the 280-ppm pre-industrial base level. As the ocean begins to absorb the excess CO_2 from fossil-fuel emissions, atmospheric CO_2 concentrations will decline, but at least 10% of the total will stay in the atmosphere for tens of thousands of years. That residual amount (roughly 55–60 ppm of the amount we have added) would keep the CO_2 concentration near 335–340 ppm (the 280-ppm baseline plus the 55- to 60-ppm residual), well above the roughly 250-ppm threshold for initiating a new glaciation (Figure 17-6).

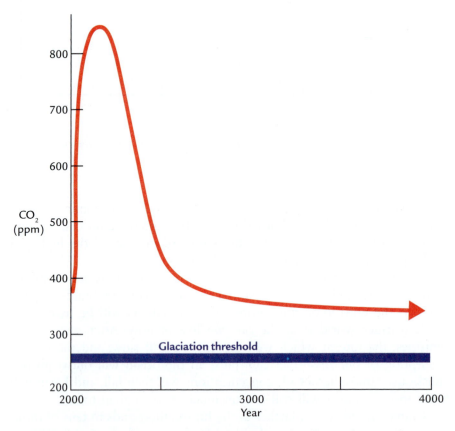

FIGURE 17-6 Atmospheric CO_2 concentrations in the distant future will decrease after most of the fossil-fuel pulse is absorbed by the ocean, but climate will not reach the threshold of glacial inception.

This calculation shows that if humans burn through a substantial part of the fossil-fuel carbon that is available and economically worth extracting, ice sheets will not appear again in the Northern Hemisphere for at least tens of thousands of years, and probably much longer than that. The residual CO_2 left in the atmosphere from our huge pulse of fossil-fuel carbon emissions will rule out the possibility of a new glacial era.

It is astonishing that we humans have now ended the 2.75-million-year history of northern hemisphere ice-age cycles for a time into the future that is beyond imagining. We have brought to a halt one of the two great systems (the other being the tropical monsoons) that have dominated climate changes during the last several million years. And we have already returned this important component of Earth's climate to a state more like that of many millions of years ago.

As noted earlier, planet Earth had been drifting very slowly toward a colder world over tens of millions of years, and we have no evidence that the natural causes behind this extremely gradual trend have stopped functioning. In the Northern Hemisphere, this slow cooling, interacting with cyclic changes in Earth's orbit, gradually produced longer and deeper cycles of glaciation during the last 2.75 million years. But now humans have overridden Nature, first by agricultural emissions of greenhouse gases in pre-industrial times and then by our greater fossil-fuel emissions during the Industrial Era.

Climate Engineering

One possibility under consideration for Earth's climatic future is **climate engineering**—the intentional manipulation of climate. We have been unintentionally engineering climatic change for millennia through our greenhouse-gas emissions, but we have only become conscious of this fact in recent decades. Technology will advance enormously in future decades and centuries, and it is likely that we will gain the ability to interfere purposefully in the operation of the climate system.

Discussions about this possibility have begun among scientists, most of whom are concerned that these schemes could have unforeseen side effects. Some scientists disapprove of even discussing climate engineering because doing so might undermine the public will to act on emissions reductions. In contrast, other scientists are quite receptive to the need for an open conversation on this issue.

Much of the climate-engineering discussion centers on minimizing the size of the oncoming CO_2 peak and the warming it will cause over the next few centuries. If we find that we can somehow counter the warming effect of the CO_2 peak at reasonable cost, the question will inevitably arise: "What climate do we want?" Because "we" means all of humanity, this discussion will be unimaginably complicated. Will a solution require an international agreement? And will all nations agree to it? Or will rogue nations try to change climate on their own? And if they do, will other nations take countermeasures? Will there be "climate wars"?

Summary

Human activities, both agricultural and industrial, have ended a sequence of northern glacial cycles that has lasted for some 2.75 million years. Ice sheets will not again appear in Canada or Scandinavia until the pulse of excess CO_2 we have put in the atmosphere is entirely absorbed by the ocean, tens of thousands of years or more into the future. In the immediate future, humanity faces the problem of adapting to a planet that will be much warmer than it has been in millions of years.

Additional Resources

Archer, D. *The Long Thaw: How Humans Are Changing the Next 100,000 Years of Earth's Climate*. Princeton: Princeton University Press, 2008.

Archer, D., and A. Ganopolski. "A Movable Trigger: Fossil Fuel CO_2 and the Onset of the Next Glaciation." *Geochemistry, Geophysics, Geosystems* 6 (2005) QO5003. doi:10.1029/2004GC000891.

Ruddiman, W. F. *Earth's Climate: Past and Future*. Chapters 6, 9. New York: W. H. Freeman, 2007.

Part 5 Summary

The centerpiece of the early anthropogenic hypothesis is the proposal that natural greenhouse-gas concentrations would by now be considerably lower than those measured in ice cores for late pre-industrial times (see Parts 1 through 3). Instead of concentrations of 285 ppm for CO_2 and 790 ppb for CH_4 at the start of the Industrial Era, the early anthropogenic hypothesis calls for natural concentrations of about 245 ppm for CO_2, and about 445 ppb for CH_4. These lower greenhouse-gas concentrations would have produced a cooler climate.

A corollary feature of the early anthropogenic hypothesis is the claim that, absent any human-generated greenhouse-gas emissions, some far-northern regions would by now have entered the early stages of a glacial world, with permanent snowfields coalescing into growing ice sheets. Several recent experiments using high-resolution climate models support this proposal. According to these simulations, a cooling of several degrees Celsius would have allowed permanent snowfields to appear in the northern Canadian Rockies, the Canadian Arctic, and

Arctic regions of northern Eurasia, covering a combined area larger than modern-day Greenland (see Chapter 15, Figure 15-7).

This evidence needs to be weighed against another anthropogenic change that has acted in the opposite direction. Forest clearance can affect climate at middle and high latitudes by increasing the reflection of incoming solar radiation and cooling nearby regions. Model experiments show that early forest clearance would have had a net cooling effect measured in tenths of a degree Celsius, or about 10 to 15% the size of the warming caused by the early anthropogenic greenhouse gases.

Given this evidence from models, it seems unlikely that changes in land-surface albedo could have canceled the appearance of permanent snow cover caused by lower greenhouse-gas levels. In the future, model simulations should test the combined effects of these competing anthropogenic factors.

The results from model simulations imply that greenhouse-gas emissions from early farming stopped the natural onset of a new glaciation within the last few thousand years. Now, the acceleration of Industrial Era gas emissions has ended the possibility of any future glaciation for tens of thousands of years or more. Given that some 50 glacial cycles in North America and Eurasia occurred during the last 2.75 million years, the cessation of these cycles for the indefinite future is a striking demonstration of human intervention in Earth's climatic history.

Small Steps Back Toward an Ice Age

T he first five parts of this book focused on several important changes during the last several thousand years: the gradual increase of CO_2 and CH_4 concentrations; the relative contributions of humans and natural factors to those increases; and the effect of the rising greenhouse-gas levels in countering part of a natural longer-term cooling.

The natural long-term cooling continued during recent centuries until the start of the Industrial Era. Following a relatively warm interval late in the Medieval Era, often referred to as the **Medieval Warm Period,** a small cooling culminated in an interval known as the **Little Ice Age** (Chapter 18). Paleoclimatic reconstructions place the start of the Little Ice Age sometime between the years 1000 and 1400, and the period ends around 1850. During this cooling, sea ice advanced southward near Iceland, and **mountain glaciers** descended to lower elevations in the Alps and in Scandinavia. These changes are thought to have been driven largely by natural factors: decreasing summer insolation at high northern latitudes, volcanic explosions that blocked some incoming solar radiation, and natural variations in the strength of the Sun.

Near 2,000 years ago, ice-core records show that the gradually rising CO_2 trend began to flatten out (Figure 6i-1 top), and high-resolution measurements show several distinct drops in CO_2 values (Figure 6i-1 bottom). One drop of about 5 ppm occurred between the years 200 and 600, and a larger decrease of about 10 ppm took place between 1200 and 1625.

The prevailing scientific view is that natural factors in the climate system were responsible for these abrupt CO_2 drops, but explanations that call only on natural factors have a flaw (Chapter 19). Models that simulate the complex interactions between Earth's climate system and its carbon reservoirs suggest that natural cooling processes cannot explain most of the CO_2 drop between 1525 and 1600. Similarly, natural factors do not appear to account for several short-term decreases in atmospheric CH_4 concentrations during the same period. Perhaps humans were an important factor in these greenhouse-gas drops. One possibility is that calamities that killed enormous numbers of people during spans of decades to a few centuries caused the gas concentrations to fall.

Chapter 20 explores historical records of two kinds of calamities that killed tens of millions of humans (about 10–20% of the global population) during the same intervals as the CO_2 and CH_4 decreases. One factor was the spread of massive **pandemics**—outbreaks of diseases that simultaneously afflicted people across several continents. The other

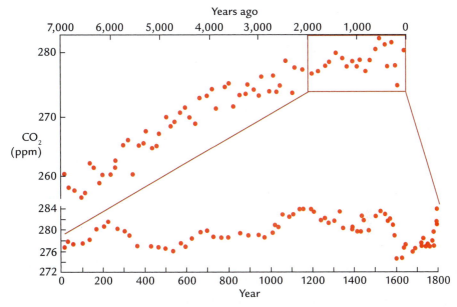

FIGURE 6i-1 The high-resolution CO_2 trend at Law Dome, Antarctica, covering the last 2,000 years (bottom) shows several short-term dips that reverse the long-term rise at Dome C, Antarctica, since 7,000 years ago (top). *[Law Dome CO_2 trend from C. McFarling Meure et al., "Law Dome CO_2, CH_4 and N_2O Ice Core Records Extended to 2000 Years BP," Geophysical Research Letters 33 (2006): L14810, doi:10.1029/2006GL026152. Dome C CO_2 trend from EPICA Community Members, "Eight Glacial Cycles from an Antarctic Ice Core," Nature 429: 623–628.]*

factor was civil strife—breakdowns of existing social and economic structures caused by wars and their immediate aftermath.

These episodes of mass human mortality could have disrupted the operation of the carbon system in several ways (Chapter 21). Reforestation of previously cleared areas would have occurred as the deaths of tens of millions of people left vast areas of previously farmed land abandoned. This regrowth of forests on what had once been cleared land would likely have taken about a century. The storage of carbon in the wood of these regrowing forests would have reduced CO_2 concentration in the atmosphere. Reductions in methane emissions (and in methane's concentration in the atmosphere) are likely to have occurred during times when the infrastructure used to irrigate rice paddies was torn up by civil strife, when millions of livestock died, and when biomass burning was reduced. Chapter 21 also evaluates the likelihood that anthropogenic forest clearance at high northern latitudes late in the pre-industrial era caused an albedo-related cooling in that region.

CHAPTER 18

THE LITTLE ICE AGE

Compared to a full ice age, the Little Ice Age was a minor episode, with an average global cooling of just a few tenths of one degree Celsius from the slightly warmer temperatures that preceded it. In comparison, the typical cooling from full interglacial climates into full glaciations is more than 10 times as large—5°C or more. The Little Ice Age was also brief in duration, starting and ending in a little over 500 years, compared to the gradual buildup of full ice ages that took tens of thousands of years. From a long-term climatic perspective, the Little Ice Age was not that big a deal.

Still, to people living in vulnerable regions, it was no trivial matter. In cold areas near the margins of high-latitude sea ice and ice caps, as well as high-altitude mountain glaciers in middle latitudes, snow and ice advances were major changes. In some of these places, subsistence farming, which had already been marginal, became impossible.

Observations of the Little Ice Age

Because this cool interval occurred recently, it has been well documented in **historical archives**. One useful record is the number of days each winter that the fishing fleet in northern Iceland was unable to go to sea because sea ice jammed the ports (Figure 18-1). Winter ice had been rare between the years 1000 and 1200, but it became much more frequent by 1600. Because sea ice builds up and retreats relatively slowly (over many years or even decades) in response to changes in temperature, it is a good indicator of the climate of the high-latitude North Atlantic around Iceland.

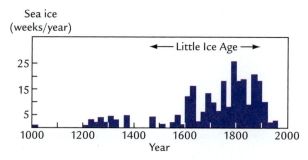

FIGURE 18-1 Changes in frequency of impassable sea ice off the coast of Iceland during the last 1,000 years. *[Adapted from H. H. Lamb,* Climate—Past, Present, and Future. *Vol. 2 (London: Methuen, 1977).]*

During the relative warmth of the late Medieval Era, a Norse colony established on Greenland by Eric the Red grew wheat on farms in the ice-free land around the southern margins of the ice sheet. Later, during the Little Ice Age, the Greenland colony perished for reasons that still remain unclear. As climate cooled, local food sources declined, and trade with Europe was choked off by the increase in sea ice. In addition, conflict had begun between the Norse settlers and the native Inuit people living farther north whose lives were partly dependent on the presence of sea ice. At times during the Little Ice Age, people as far south as northern Scotland saw Inuit people in kayaks fishing along a sea ice border that had expanded well beyond its normal limit in the ocean between Greenland and Norway.

Advances and retreats of mountain glaciers and small **ice caps** provide another useful index of regional temperature trends, again because they respond to changes that occur over years to decades. Today, thousands of valley glaciers exist in high-mountain areas at many latitudes, and numerous ice caps cover flatter terrain on high-Arctic islands (Figure 18-2). Almost all of these bodies of ice are now melting in the rapidly rising warmth of the Industrial Era, but during the cooler intervals of the Little Ice Age, most of them grew to sizes well beyond their current extent.

In the eastern Canadian Arctic, ice caps exist across a large region west of Greenland. Baffin Island, a large island the size of California off the northeast coast of mainland Canada, provides a glimpse of what Little Ice Age climate was like in the Arctic (see Figure 18-2). Several ice caps still remain along the higher parts of the island, surrounded by curious halo-like features with dead **lichen** (Figure 18-3). Lichen are small, primitive vegetation forms that grow by extracting nutrients from minerals in rocks.

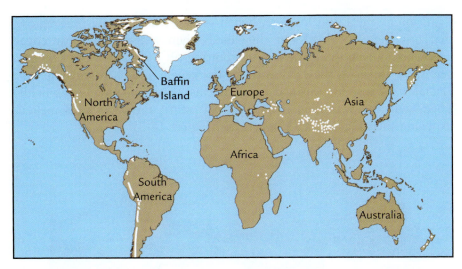

FIGURE 18-2 Schematic representation of the tens of thousands of glaciers and ice caps (in *white*) today. [*Adapted from* National Geographic World, *No. 18, February 1977).*]

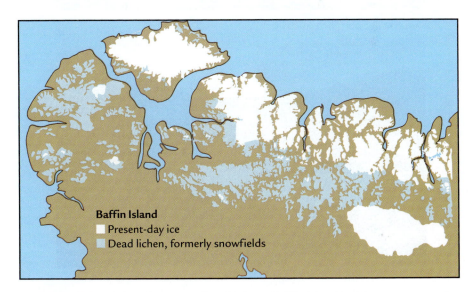

FIGURE 18-3 Modern ice cover (in *white*) on Baffin Island, with areas covered by permanent snow during the Little Ice Age (in *blue*). [*Adapted from J. T. Andrews et al., "The Laurentide Ice Sheet: Problems of the Mode and Speed of Inception,"* WMO/IMAP Symposium on Long-term Climatic Fluctuations *421(1975).*]

Prior to the Little Ice Age, lichen grew in these snow-free regions that now have "halos," but they died when year-round snow cover developed during the Little Ice Age, cutting off their supply of sunlight. The snow cover in these areas never thickened enough to develop into glacial ice caps, but instead remained as semipermanent snowfields. Then, with the warming of the last 150 years, the snowfields melted, exposing the areas of dead lichen (see Figure 18-3).

A

B

FIGURE 18-4 Photographs of Thunderbird Glacier in Glacier National Park, Montana, taken from the same vantage point—one in 1907, just after the end of the Little Ice Age (A), and the other in 2007, with little ice left (B). *[A: Morton Elrod, courtesy of GNP Archives. B: Dan Fagre/Greg Pederson/USGS.]*

Farther south, at latitudes warmer than the Arctic, some mountain regions are high enough (and cold enough) that they sustained glacial ice during the slightly cooler Little Ice Age. Glaciers descended farther down mountain valleys in the western U.S. Rockies (Figure 18-4), the central European Alps, the Himalaya and the Tibetan Plateau of south-central Asia, and the Andes.

At times, mountain glaciers advanced far down into valleys in the Alps of Switzerland, France, Austria, and Italy, and even overran farmhouses and villages. Near these glaciers, agriculture became difficult or impossible as the length of the frost-free season shrank. Crops were also more difficult to grow along their natural northern limits, and some kinds of grapes that grew in England in medieval times before 1200 could no longer be grown during the Little Ice Age.

The largest regional cooling trends during this climatic deterioration probably averaged no more than 1°C (less than 2°F), but even a cooling that small can matter in vulnerable regions. Dropping the temperature from 1°C (almost 34°F) to 0°C (32°F) can turn rain to snow, and a similarly small cooling past the freezing threshold can cover the surface of oceans with ice. In already frigid regions, at high latitudes and high altitudes, the Little Ice Age cooling was a major event that altered the lives of those few people living there.

In contrast, at lower and middle latitudes well away from very cold regions, with little or no snow and ice available to amplify climatic change, the cooling was much smaller, just a few tenths of 1°C. In hotter tropical climates, a change from 30°C (86°F) to 29.5°C (85°F) might even have felt like a slight change for the better, assuming that people even sensed it happening.

Climate Proxies and Reconstructed Temperature Changes

To assess the Little Ice Age, scientists have attempted to quantify the amount of cooling after the end of the late Medieval Era. Because thermometers did not yet exist for most of this time, investigators have been forced to rely on **climate proxies**—responses registered in natural archives of climate. Two of the major proxies investigated for this purpose are **tree rings** and ice cores, each of which contain annual layers that can be used to date the embedded climatic records (Figure 18-5).

Each of these proxies records past temperatures in its own way: tree rings display changes in width or density (they have wider, less dense rings in years with warm summers), and ice cores record oxygen-isotope

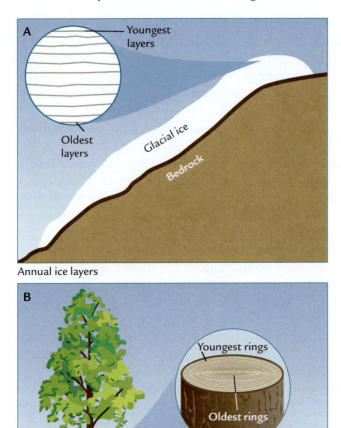

FIGURE 18-5 Two climatic archives with annually deposited layers that can be used to date the climatic records they contain: glacial ice (A) and trees (B).

indices in their frozen annual layers. The part of each proxy record that spans the last several decades is used to establish a quantitative link to temperature changes measured by thermometers at nearby meteorological stations. This relationship is then projected back to the earlier part of the record, using well-dated proxy measurements to estimate temperature.

Temperature is not the only factor that affects climate proxy responses. For example, tree rings also respond to precipitation, especially in relatively dry regions, and the oxygen-isotope index of glaciers and ice sheets responds to factors such as the time of year the snow falls and changes in the sources of the water vapor that falls as snow and turns to ice. To overcome these complications, the methods of proxy analysis attempt to isolate the unique effects caused by past temperature changes.

Dozens of well-dated proxy records from many regions are compiled and integrated to produce a single best estimate of temperature change. Most estimates of temperature change have been made for the Northern Hemisphere, where most proxy records are located. At this point, useful records are scarce in the ocean-dominated Southern Hemisphere.

Because of the varying methods of analysis and different selections of proxy records, reconstructions of past northern hemisphere temperatures differ considerably (Figure 18-6). All of them show a net cooling between the medieval years 1000–1200 and the middle of the Little Ice Age in the 1600s, but the amplitude of the cooling varies from as little as 0.2°C in some records to as much as 0.5°C in others, with an average around 0.3° to 0.35°C (about 0.55° to 0.65°F). Those reconstructions that show a larger cooling tend to rely more on proxy sites from high northern latitudes where climatic changes are larger than the hemispheric average because of amplification by snow and ice feedback.

All of the proxy records, as well as the thermometer measurements of the last 150 years, show a warming since 1850 that is unprecedented

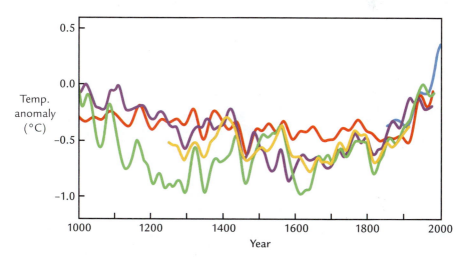

FIGURE 18-6 Several reconstructions (in *various colors*) of changes in northern hemisphere temperature during the last 1,000 years. *[Adapted from P. D. Jones and M. E. Mann, "Climate Over Past Millennia,"* Reviews of Geophysics *42 (2004), No. 2, RG2002, doi:10.1029/2003RG000143.]*

in size and rate compared to the previous eight centuries. Most scientists attribute this warming to the rapid Industrial Era increase in greenhouse gases that began in 1850 (see Chapter 17).

Natural Explanations for the Little Ice Age Cooling

The longer-term cooling trend between 1000–1200 and 1600 is widely thought to be natural in origin. One likely explanation for the cooling at high latitudes is the steady decrease in northern hemisphere summer insolation over the last 10,000 years (recall Chapter 1, Figure 1-10). The paleoclimate scientist Darryl Kaufman and colleagues recently used proxy indicators to reconstruct estimated temperature in the Arctic during the last 2,000 years and found a slow persistent cooling (Figure 18-7). They noted that this trend tracks the slow decrease in summer insolation at high northern latitudes during the same interval.

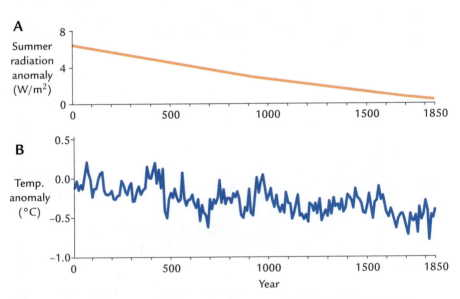

FIGURE 18-7 Decreases in summer half-year insolation at 65°N (A) and in estimated Arctic temperatures (B) during the last 2,000 years. *[(A) adapted from A. L. Berger, "Long-term Variations of Caloric Insolation Resulting from the Earth's Orbital Elements," Quaternary Research 9 (1978): 139–167; and (B) from D. S. Kauffman et al., "Recent Warming Reverses Long-term Arctic Cooling," Science 325 (2009), 1236–1239, doi:10.1126/science.1173983.]*

FIGURE 18-8 The 1991 eruption of Mount Pinatubo in the Philippines cooled global climate for the following year. *[Time & Life Pictures/Getty Images.]*

Volcanic eruptions are thought to be responsible for some of the short-term cooling episodes that lasted for intervals of years to a decade or more. Eruptions throw sulfurous gases high into the **stratosphere** and produce sulfate aerosol particles that block some of the incoming solar radiation, thereby cooling climate. The 1991 eruption of Mount Pinatubo in the Philippines (Figure 18-8) confirmed that volcanic eruptions of moderate size can cool climate for a year or two.

Larger eruptions known to have occurred during the Little Ice Age probably cooled global climate for at least that long, and a closely spaced series of eruptions could have caused cooler intervals lasting a decade or more. Clusters of large eruptions that reduced incoming solar radiation occurred during several years when (reconstructed) northern hemisphere temperatures briefly fell, such as 1180, 1250, 1450, 1600, and 1810 (Figure 18-9).

Changes in the amount of radiation emitted from the Sun have also been considered as a possible cause of temperature changes on Earth. Satellite observations of incoming solar radiation covering the last 30 years show cyclic (11-year) variations of about 0.1% around a mean

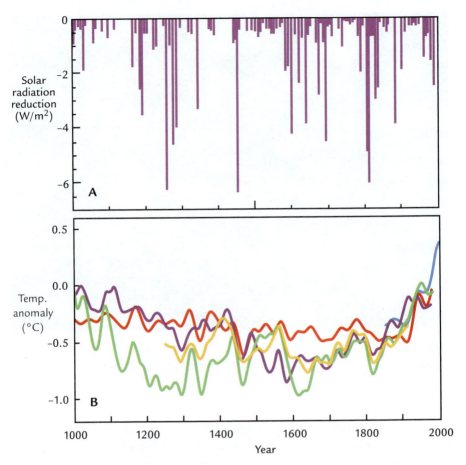

FIGURE 18-9 During the last 1,000 years, the reductions of incoming solar radiation caused by volcanic eruptions (A) account for some of the brief drops in reconstructed northern hemisphere temperature (B). *[Volcanic record adapted from T. Crowley et al., "Modeling Ocean Heat Content Changes During the Last Millennium,"* Geophysical Research Letters *30 (2003), doi:10.1029/2003GL017801. Temperature reconstruction adapted from P. D. Jones and M. E. Mann, "Climate Over Past Millennia,"* Reviews of Geophysics *42 (2004), No. 2, RG2002, doi:10.1029/2003RG000143.]*

value that has held steady through that time (Figure 18-10). These cycles are linked to processes acting within the Sun that produce **sunspots,** which darken part of the Sun's surface. However, during periods of high sunspot activity, overall solar emissions increase. These oscillations in solar radiation have caused a small 11-year global temperature oscillation of about 0.1°C.

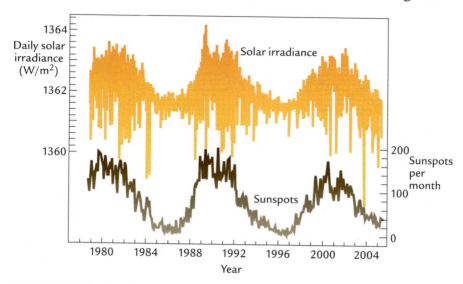

FIGURE 18-10   Satellite measurements of solar radiation arriving at the top of the atmosphere between 1978 and 2004, compared to sunspot numbers. *[Solar radiation values adapted from the World Radiation Center (WRC) at www.pmodwrc.ch/, and sunspot data from the Solar Influences Data Analysis Center (SIDC) at sidc.oma.be/index.php3, both accessed December 18, 2012.]*

Until the early 2000s, scientists thought that Sun-like stars underwent larger changes in strength over decades to centuries, but closer investigations have not fully supported this idea. While solar variations during the last 150 years were thought to explain as much as 20% to 33% of the observed Industrial Era warming since 1850, they are now credited with at most 5% to 10% of the total. Researchers have now extended this view into earlier (pre-industrial) times. From the currently available evidence, changes in Sun strength do not appear to have been a large factor in the gradual background cooling into the Little Ice Age.

Summary

The long-term cooling that began thousands of years ago (see Chapter 15) continued during the last millennium, until the abrupt reversal to a rapid warming since 1850. At a global scale, the size of the cooling from the late Medieval Era to the Little Ice Age amounted to only a few tenths of a degree Celsius, but the larger cooling that occurred at high northern latitudes led to advances in sea ice, expansion of Arctic snowfields, and descent of mountain glaciers. The primary natural factors that drove this cooling were a long-term decrease in northern summer insolation, short-term volcanic eruptions, and (perhaps) small variations in the strength of the Sun.

Additional Resources

Bradley, R. S. *Paleoclimatology: Reconstructing Climates of the Quaternary: International Geophysics Series*. 2nd edition. San Diego: Academic Press, 1999.

Bradley, R. S., and P. D. Jones. *Climate Since A. D. 1500*. London: Routledge, 1992.

Grove, J. M. *The Little Ice Age*. London: Routledge, 1988.

Jones, P. D., and M. E. Mann. "Climate Over Past Millennia." *Reviews of Geophysics* 42 (2004), No. 2, RG2002. doi:10.1029/2003RG000143.

CHAPTER 19

WERE THE DROPS IN CO_2 AND CH_4 NATURAL?

Ice cores recovered from the highest summits of the Antarctic ice sheet (such as Dome C, the site of the records analyzed in Chapter 9) hold the longest records but accumulate at relatively low resolution because so little snow falls. Cores from smaller domes and ridges lower on the ice sheet receive more precipitation, and sheltered locations into which winds blow additional snow add to the rates of accumulation. These sites provide higher resolution records.

One such site is Law Dome, whose high-resolution records of CO_2 and CH_4 span recent millennia (Figure 19-1). The records of both gases show short-term drops in concentrations defined by numerous data points, an indication that the decreases do not result from random "noise" or analytical uncertainty. During intervals such as the years from 1200 to 1400 and the years from 1500 to 1700, the drops in gas concentrations more or less coincide.

Natural Explanations for CO_2 Decreases

As summarized in Chapter 18, decreasing summer insolation, volcanic explosions, and intervals of slightly diminished Sun strength are believed to be important factors in the longer-term cooling that marked

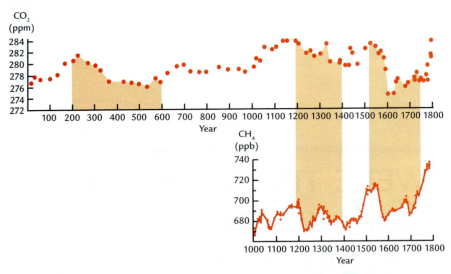

FIGURE 19-1 CO_2 and CH_4 decreases during the last 2,000 years. *[CO$_2$ record from C. McFarling Meure et al., "Law Dome CO$_2$, CH$_4$ and N$_2$O Ice Core Records Extended to 2000 Years BP," Geophysical Research Letters 33 (2006): L14810, doi:10.1029/2006GL026152. CH$_4$ record adapted from L. E. Mitchell, E. J. Brook, T. Sowers, J. R. McConnell, and K. Taylor, "Multidecadal Variability of Atmospheric Methane, 1000–1800 C.E." Journal of Geophysical Research 116 (2011), GO2007, doi:10.1029/2010JG001441.*

the Little Ice Age. Many scientists also interpret the short-term decreases in CO_2 and CH_4 shown in Figure 19-1 as having a natural origin. In this view, naturally driven cooling caused the decreases in CO_2 and CH_4. New evidence, however, suggests this view needs to be reexamined.

Temperature can exert a natural control on atmospheric CO_2 concentrations via several natural processes, one of which is the amount of CO_2 dissolved in seawater, which depends in part on ocean temperature (see Chapter 11). When ocean temperatures rise, CO_2 is emitted from the oceans into the atmosphere. But when the oceans cooled, as was the case when the Little Ice Age started, their capacity to hold dissolved CO_2 would have increased, which would have removed CO_2 from the atmosphere (Figure 19-2A).

Another link between temperature and atmospheric CO_2 concentrations operates through changes in **soil litter**—the leaves, twigs, grass, and other debris found on the ground and in the upper soil layers of forested and grassland areas (Figure 19-2B). Over intervals of years to

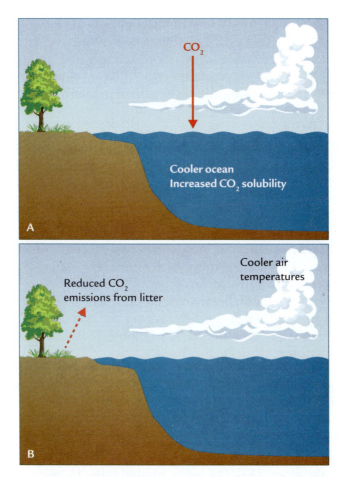

FIGURE 19-2 Two links between climatic cooling and CO_2 concentrations in the atmosphere: a cooler ocean holds more dissolved CO_2 (A); and cooler air temperatures reduce emissions of CO_2 from vegetation litter (B).

decades, the organic carbon in this debris reacts with oxygen, surrendering carbon to the atmosphere as CO_2 ($C + O_2 \rightarrow CO_2$). Some of the CO_2 release occurs quickly as leaves decompose or as fires burn through the undergrowth, but larger branches that fall to the ground may take decades to rot completely and release their carbon. When temperature drops, as it did during the Little Ice Age, these emissions from ground litter diminish, reducing the CO_2 concentration in the atmosphere.

Other natural factors work in the opposite direction, increasing atmospheric CO_2 levels as temperatures cool. When the frost-free season is shortened, vegetation grows more sparsely at high latitudes and altitudes. This net loss of vegetation would have released more carbon into the atmosphere during the cooling as Earth moved into the Little Ice Age.

A second factor comes into play when the CO_2 concentration in the atmosphere falls. Because carbon dioxide acts through the process of photosynthesis to promote growth of trees and other vegetation, an atmosphere with decreasing CO_2 levels will provide less of this **CO_2 fertilization**. With less production of vegetation on the continents and less removal of CO_2 from the air, atmospheric CO_2 concentrations will rise.

Assessing the net effect of these competing processes on atmospheric CO_2 concentrations during times of temperature change is very complicated, because all of them are at work simultaneously. As a result, scientists turn to **carbon/climate models** designed to simulate all of the interactions at once. Each model produces an estimate of the change in atmospheric CO_2 that occurs in response to specified changes in global temperature, expressed as ppm/°C.

Results from ten such models indicate that CO_2 concentrations in the atmosphere should fall by an average of about 8 ppm for each 1°C drop in global temperature (Figure 19-3). The total range of model-estimated decreases varies from 3 to 16 ppm/°C.

These simulations pose a major problem for natural explanations of the short-term drops in atmospheric CO_2, particularly the 10-ppm CO_2 decrease from 283–284 ppm in the early to middle 1500s to 273–274 ppm by the early 1600s (see Figure 19-1). For this interval, reconstructions of the amount of northern hemisphere cooling range from 0° to 0.5°C and average about 0.17°C (see Chapter 18, Figure 18-6). If the average northern hemisphere cooling of 0.17°C represents the global average, the relationship derived from the models predicts that the cooling from the 1500s to the 1600s should have caused a CO_2 drop of roughly 1.3 ppm (0.17°C × 8 ppm per °C).

This 1.3-ppm drop would amount to just a small fraction of the observed 10-ppm drop that actually occurred by the early 1600s, and of the approximately 7-ppm decrease that was sustained into the early 1700s (see Figure 19-1). To account for the full 7- to 10-ppm amplitude of the drop, the relationship between CO_2 and temperature would need to have been 35–60 ppm/°C, far above the range indicated by climate models (see Figure 19-3).

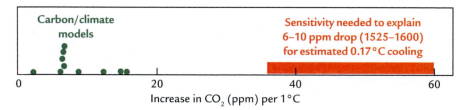

FIGURE 19-3 Model estimates of the sensitivity of atmospheric CO_2 concentrations to changes in global mean temperature. *[Adapted from D. C. Frank et al., "Ensemble Reconstruction Constraints on the Global Carbon Cycle Sensitivity to Climate," Nature 463 (2010): 527–530.]*

At present, the Southern Hemisphere has too few sites with temperature proxies to yield a reliable estimate of its contribution to the global average temperature change during the Little Ice Age, but preliminary data suggest that the cooling in the south may have been smaller than that in the north. The Southern Hemisphere is far more ocean-dominated than the Northern Hemisphere, and oceans tend to moderate climate because they are less responsive to climatic drivers than large landmasses. A smaller southern hemisphere cooling would diminish the global average cooling and even further reduce the 1.3-ppm portion of the CO_2 drop that can be explained by natural factors acting through the temperature mechanism.

These results leave scientists with two possible choices. Those who still favor a natural origin for the CO_2 drop can argue that the CO_2/temperature relationship during this one interval (from 1525 to 1600) was much larger than the mean sensitivity based on the carbon/climate models (see Figure 19-3). But then why was the normal operation of the climate system simulated by the models so different during just this one short interval?

Alternatively, if the CO_2/temperature relationship is assumed to have stayed within the range estimated by the carbon/climate models, then the relatively small cooling from 1525 to 1600 cannot explain most of the observed CO_2 drop during that interval. If this is the case, the natural explanation for the CO_2 decrease is inadequate, and a different explanation is needed.

Natural Explanations for CH_4 Decreases

Similar problems arise from investigations of short-term methane variations in pre-industrial time. A high-resolution methane record from the West Antarctic Ice Sheet Divide shows a slow net increase from around 680 ppb in the year 1000 to 730 ppb or higher by 1800, but with several distinctive short-term drops of 15–25 ppb in the early 1200s, near 1350, and in the late 1500s (see Figure 19-1). This trend closely matches a previous CH_4 signal obtained from Law Dome ice.

Several studies have compared these methane oscillations to reconstructed changes in temperature and precipitation in order to test for possible links. Temperature affects atmospheric CH_4 concentrations by warming or cooling wetlands, especially in the Arctic, and precipitation affects CH_4 levels by flooding or drying out wetlands, especially those in the tropics (see Chapter 2). In general, the methane oscillations were found to be uncorrelated, or only weakly correlated, with temperature and precipitation reconstructions from several geographic areas.

Summary

The 10-ppm CO_2 drop between 1525 and the early 1600s is the defining part of the longer-term CO_2 trend from the late Medieval Era to the Little Ice Age, but most of this abrupt decrease cannot be explained by natural causes. The same conclusion holds for most downward oscillations in a generally rising CH_4 trend during the same period. In both cases, an explanation other than natural processes is required.

Parts 1 through 4 of this book showed that greenhouse-gas emissions from farming are likely a major reason (and arguably *the* major reason) for the long-term increases in atmospheric CO_2 concentration after 7,000 years ago and in CH_4 after 5,000 years ago. If the shorter-term oscillations in the concentrations of both gases explored here are embedded in longer-term trends that are partly or mostly anthropogenic, it stands to reason that these downward oscillations may also have had a human origin.

An anthropogenic explanation for the short-term decreases in atmospheric CO_2 and CH_4 concentrations during the Little Ice Age would require mechanisms that for a time reversed the tendency of humans to emit ever-more CO_2 through deforestation and increasing CH_4 through rice irrigation, livestock tending, and biomass burning. The fact that the two largest drops of CO_2 and CH_4 within the last 1,000 years occurred during similar intervals—1200 to 1400 and 1525 to 1600—mark these times as worthy of close historical examination. Could these have been times when large-scale calamities set back the progressively growing effects humans had been having on Earth's surface and its atmospheric greenhouse-gas concentrations?

Additional Resources

Crowley, T. J. "Causes of Climate Change Over the Past 1000 Years." *Science* 289 (2000): 270–276.

Frank, D. C., J. Esper, C. C. Raible, U. Büntgen, V. Trouet, T. Stocker, and F. Joos. "Ensemble Reconstruction Constraints on the Global Carbon Cycle Sensitivity to Climate." *Nature* 463 (2010): 527–530.

Gerber, S., F. Joos, P. Brügger, T. Stocker, M. E. Mann, S. Sitch, and M. Scholtze. "Constraining Temperature Variations over the Last Millennium by Comparing Simulated and Observed Atmospheric CO₂." *Climate Dynamics* 20 (2003): 281–299.

Mitchell, L. E., E. J. Brook, T. Sowers, J. R. McConnell, and K. Taylor. "Multidecadal Variability of Atmospheric Methane, 1000–1800 C.E." *Journal of Geophysical Research* 116 (2011), GO2007. doi:10.1029/2010JG001441.

Ruddiman, W. F. *Earth's Climate: Past and Future.* 2nd ed. Chapter 16. New York: W. F. Freeman, 2007.

CHAPTER 20

MASS HUMAN MORTALITY AND CO$_2$ DECREASES

With natural processes unable to explain the abrupt CO$_2$ and CH$_4$ decreases during the last 2,000 years, could there be an anthropogenic explanation—one somehow tied to the fate of humans? Fortunately, enough population data exist for several key regions to address this question. Historical records in Europe and China extend back slightly more than 2,000 years. In addition, estimates of populations in the Americas just prior to and after the arrival of Europeans during the late 1400s and early to middle 1500s are becoming clearer.

During three intervals within the last 2,000 years, tens of millions of humans died at levels far above normal rates of mortality, and during each of those times CO$_2$ levels fell, as did CH$_4$ concentrations during several intervals after the year 1000 (Figure 20-1). The three periods of excess mortality spanned the era of the late Roman Empire (200–600), part of the Medieval Era (1200–1400), and the European conquest of the Americas (1525–1750). Although correlation does not prove causality, the clear correlation between the CO$_2$ and CH$_4$ decreases and these intervals of mass mortality suggests a cause-and-effect relationship.

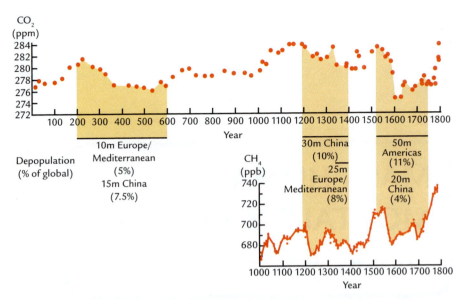

FIGURE 20-1 CO_2 and CH_4 decreases during the last 2,000 years correlate with intervals of massive population loss. [CO_2 record from C. McFarling Meure et al., "Law Dome CO_2, CH_4 and N_2O Ice Core Records Extended to 2000 Years BP," Geophysical Research Letters 33 (2006): L14810, doi:10.1029/2006GL026152. CH_4 record adapted from L. E. Mitchell, E. J. Brook, T. Sowers, J. R. McConnell, and K. Taylor, "Multidecadal Variability of Atmospheric Methane, 1000–1800 C.E." Journal of Geophysical Research 116 (2011), GO2007, doi:10.1029/2010JG001441. Population levels from C. McEvedy and R. Jones, Atlas of World Population History (New York: Penguin Books, 1978); and from W. M. Denevan, "The Pristine Myth: The Landscape of the Americas in 1492," Annals of the Association of American Geographers 82 (1992): 369–385.]

Europe and the Mediterranean: Bubonic Plague

Twice in the pre-industrial era, European civilizations bordering the Mediterranean Sea were devastated by outbreaks of bubonic plague that killed a third or more of the inhabitants within a few years (Figure 20-2). In both cases, the high mortality led to a partial social breakdown, described in William McNeill's *Plagues and Peoples*. Terrified people living in cities, towns, and villages threw the decomposing bodies of their dead relatives and friends out into the street and huddled in their houses, hoping to be spared. Farmers living outside settlements were also vulnerable to these outbreaks, and their deaths left many farms abandoned. Historical records tell us that these abandoned lands reverted to the wild, slowly becoming covered with natural vegetation.

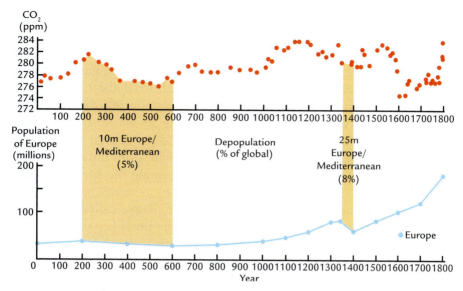

FIGURE 20-2 Changes in CO_2 concentration and European population during the last 2,000 years. [*CO_2 record from C. McFarling Meure et al., "Law Dome CO_2, CH_4 and N_2O Ice Core Records Extended to 2000 Years BP,"* Geophysical Research Letters 33 (2006): L14810, doi:10.1029/2006GL026152. *Population levels from C. McEvedy and R. Jones,* Atlas of World Population History *(New York: Penguin Books, 1978).*]

By the year 200, the vast Roman Empire extended from the Middle East to North Africa and Spain and northward along the Atlantic coast to Britain. The Mediterranean Sea was the central "highway" during this highly commercial era, and ships carried lumber, food, and slaves from colonies throughout the empire. Unfortunately, this avenue of commerce was also a highway for disease.

The population of Roman Europe reached a peak at, or just before, the year 200, but then it began a gradual decline that continued until 600. Historians have long argued about the causes of this population loss. Hypotheses that have been considered include wars, invasions by hostile people from the north, Roman mismanagement, or some other breakdown of central authority. Some historians believe that disease played an important role.

The water brought to Rome in aqueducts from mountain sources was remarkably pure, so waterborne disease was not the problem. Rome had experienced outbreaks of malaria (until then endemic to Africa) in 79 and 125, probably linked to the buildup of the nearby Pontine marshes by eroded river silts from the Tiber River. But these outbreaks were local in extent and small in size. The Antonine Plague (also called Galen's Plague) struck between the years 165 and 189 and resulted in a somewhat higher mortality. The disease responsible for this event is not certain, but it may have been smallpox.

During the next few centuries, mortality from disease increased. The Plague of Cyprian (also possibly smallpox) afflicted the Empire from Egypt in the east to Scotland in the northwest during the year 250; historians again describe large areas reverting to waste. By the 500s, the Western Roman Empire had collapsed, and the capital of the Eastern (Byzantine) Empire had been established in Constantinople (now Istanbul), a seaport along the narrow straits joining the Mediterranean and Black Seas.

In 540, an enormous outbreak of bubonic plague called the Plague of Justinian (after the Byzantine emperor at the time) began. The earliest signs appeared in the eastern Mediterranean (Egypt), from a source suspected to lie farther to the east, perhaps in semi-arid central Asia. The plague spread rapidly westward on ships carrying rats that hosted the flea *Yersinia pestis* (Figure 20-3), and it moved inland from port cities as rats infiltrated the countryside. Within a few years, bubonic plague had swept across most of the Byzantine Empire.

FIGURE 20-3 Rats use tie lines to move from ships into seaport cities. *[Stephen Dalton/Photoshot.]*

Bubonic plague caused a horribly painful death: lymph nodes in the armpits and groin swelled into bulbous protrusions called "buboes." This swelling was followed by fever, thirst, coma, convulsions, vomiting, and diarrhea. Just before death, which usually came within two weeks, black spots appeared on the bodies of the victims. Historical accounts tell of deserted streets in towns and cities piled with decaying bodies that no one would attend to. More than 25% of the people in Europe died. Entire towns and villages were wiped out and farms were abandoned. Grapes withered on the vine, and unharvested grain was left to rot on the ground. Smaller outbreaks continued every decade or two until 590, decimating each new generation.

The Plague of Justinian capped a cumulative drop of the European population from about 36 million people in the year 200 to about 26 million by 600, a total decrease of almost 40%. The decrease in concentration of atmospheric CO_2 from 281 ppm to 276 ppm during this interval suggests a possible link between mortality and CO_2 levels (see Figure 20-2).

The next 700 years saw a relative lull in the incidence of plague in the Mediterranean region. Sporadic small-scale outbreaks occurred in parts of Europe, and larger outbreaks struck Islamic countries farther to the east during the 600s and 700s. European populations (and global atmospheric CO_2 levels) rebounded slowly from 600 to 1000, although populations in some areas didn't return to pre-plague levels for many centuries. Then, between 1000 and 1300, the population of Europe surged to about 75 million people. Trade increased, cities grew, and major cathedrals were built. CO_2 concentrations in the atmosphere also rose rapidly, reaching values as high as 283–284 ppm between 1000 and 1200.

Between 1347 and 1353, a second wave of bubonic plague struck, again probably starting somewhere in central Asia, moving west through eastern Mediterranean countries, and sweeping northwest across Europe. Within five years, one-third of the population of Europe, some 25 million people, died. This pandemic came in a much deadlier form: it spread simply by people coughing. Again, blackened spots called "tokens" appeared on the skin of the victims, giving this pandemic its name: the Black Death. Albrecht Dürer captured the morbid spirit of this era in his lithograph "The Four Horsemen of the Apocalypse," portraying pestilence as a skeletal rider (the "pale horseman") on an emaciated horse (Figure 20-4).

Once again, as societal norms broke down, bodies were left to rot in streets, and many survivors lost hope for the future, some turning to

FIGURE 20-4 *The Four Horsemen of the Apocalypse,* a woodcut by Albrecht Dürer. *[Private Collection/The Bridgeman Art Library.]*

hedonistic behavior, and some making scapegoats of imagined witches as an outlet for their fears and frustrations. Again, entire country villages disappeared, by one estimate perhaps 25% of those on the European continent. As farmers died, crops were again left unharvested in the fields. During this pandemic, however, population levels in the southern, western, and central parts of Europe were much higher than they had been during Roman times, and many farms that were initially abandoned were probably soon reoccupied by family members, neighbors, or others. In less heavily populated northern and northeastern Europe, fewer farms were reoccupied and many reverted to the wild. Outbreaks of plague continued for another 50 years, "cropping" vulnerable members of each new generation. Total mortality during the late 1300s was close to 33 million people, or 44% of the pre-plague population.

The Black Death pandemic falls within a noisy CO_2 decline that began near 1200 and continued into the 1600s, but the pandemic interval is not marked by an obvious CO_2 drop. (The middle 1300s were also marked by a short-lived decrease in CH_4 concentrations [see Figure 20-1]).

The last significant plague outbreak occurred in the early 1700s, even though a vaccine was not developed until 1884. Putting afflicted cities under quarantine (described in Albert Camus' novel *The Plague*) helped to prevent its spread. Other diseases were locally epidemic in Europe during pre-industrial and industrial times—among them small-pox, typhus, cholera, and yellow fever—but these outbreaks never killed people at the level of the two bubonic plague pandemics.

War played only a minor role in Europe's pre-industrial population trend (see Figure 20-2). Roughly 8 million people died in Germany and Belgium during the Thirty Years War from 1618 to 1648, but they amounted to less than 10% of the more than 100 million people in Europe at the time, and less than 2% of the global population. The conflict slowed the rate of European population growth from 1600 to 1700, but did not reverse it.

Famine is often cited as having an important effect on human pop-ulations, but in Europe it was not much of a factor in pre-industrial mortality episodes. The worst pre-industrial European famine occurred between 1315 and 1317 as a result of several cold wet years that pro-duced small harvests. Yet within a decade, the population had recovered from the loss of these few million people. In Europe, disease (chiefly bubonic plague) was by far the main cause of mass mortality.

China: Civil Strife

The population of China since at least 2,000 years ago is reasonably well known from sporadic census counts (Figure 20-5). Through the early part of that interval, the population trend was remarkably similar to that of Europe, with an early peak near the year 200, slow decreases from 200 to 600, a slow recovery from 600 to 1000, and rapid growth after 1000. Subsequently, the trend in China diverged from the one in Europe, with a population decrease in China from 1200 to 1400, and very rapid growth thereafter.

Despite the similarities with the trends in Europe, most historians do not see disease as the primary cause of the episodic population decreases in China. Rather, they point to periods of intense civil strife,

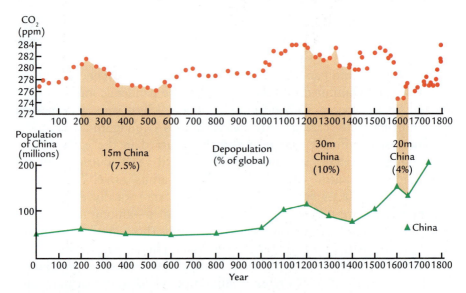

FIGURE 20-5 Changes in CO_2 concentration and the population of China during the last 2,000 years. [CO_2 record from C. McFarling Meure et al., "Law Dome CO_2, CH_4 and N_2O Ice Core Records Extended to 2000 Years BP," Geophysical Research Letters 33 (2006): L14810, doi:10.1029/2006GL026152. Population levels from C. McEvedy and R. Jones, Atlas of World Population History (New York: Penguin Books, 1978).]

mostly resulting from invasions by central Asian peoples from across the northwest border. These incursions are thought to have been underway by at least 2,500 years ago, and are responsible for the early construction of the Great Wall of China as a defensive measure.

By 2,000 years ago, China had 50–60 million people, after which census counts document two major population drops. Between 200 and 600, roughly 15 million people (7.5% of the global population) died. Most historians favor civil strife ("agrarian unrest") caused by invasions from the northwest as the main reason for this drop. Part of the apparent population loss may actually have been an artifact of people fleeing to more remote parts of southern China to escape unrest in the north and consequently dropping out of subsequent census counts. The population drop from 200 to 600 coincides with the 5-ppm CO_2 decrease from 281 ppm to 276 ppm (see Figure 20-5).

The second major population decrease (from 1200 to 1400) resulted from incursions by horse-mounted Mongol warriors from the northwest. Although relatively small in numbers, these central Asian people were amazingly effective warriors, because they had developed new stirrups and saddles that permitted a high degree of maneuverability on horseback. The invasions began in the early 1200s and reached a culmination with the arrival in 1278 of Genghis Khan and his kin (Figure 20-6). The Khans established the Yuan dynasty in northern China, causing millions of people to flee to southern China, with some of them again probably falling off the official census counts for the Chinese Empire. Historians estimate the possible range of mortality between 1200 and 1400 at 25–50 million people, with a conservative value of 30 million. Those 30 million people could not have been slain one-by-one with swords or knives. Instead, many of the deaths resulted from starvation after the Khans destroyed much of the agricultural infrastructure: irrigation ditches that carried water to rice paddies, and roads and canals used to transport food to cities.

This population drop coincides with the 5-ppm decrease in atmospheric CO$_2$ from 284 ppm to 279 ppm between 1200 and 1400 (see Figure 20-5). The deaths in China combined with those from the Black Death in Europe amounted to almost 20% of the world population of the time. In addition, the CH$_4$ concentration fell by almost 30 ppb in the middle 1200s, although it largely recovered in subsequent decades (see Figure 20-1).

FIGURE 20-6 Drawing of Genghis Khan. [Ms Pers.113 f.29 Genghis Khan (c.1162–1227) Fighting the Tartars, from a book by Rashid-al-Din (1247–1318) (gouache), Persian School, (14th century)/Bibliotheque Nationale, Paris, France/The Bridgeman Art Library.]

Another interval of high mortality—20 million deaths—occurred in the early 1600s during the Manchu conquest of the Ming dynasty, accompanied by a smallpox epidemic. This population drop occurred near the end of an interval between 1525 and the early 1600s when CO_2 values fell by almost 10 ppm and CH_4 values by about 30 ppb (see Figures 20-1 and 20-5).

The population history of the rest of southern Asia is poorly known for most of the pre-industrial era. Records in India are scant prior to the Mughal Empire in the 1500s. No comprehensive records have been published for the rest of Southeast Asia.

The Americas: The Virgin Soil Pandemic

Aside from professional anthropologists and archeologists, and readers of Charles Mann's *1491* and Jared Diamond's *Guns, Germs, and Steel*, most people today remain unaware of the worst pre-industrial disaster ever to befall the occupants of any continent—the near-total eradication of the indigenous peoples of the Americas by disease between 1492 and the 1700s (Figure 20-7). According to the anthropologist William Denevan, an estimated 55–60 million people lived in the Americas prior to European arrival. By 1700, a host of diseases introduced by Europeans and their livestock had reduced that number by 85–90% to some 6–7 million survivors. The environmental historian Alfred Crosby called this great annihilation the **virgin soil pandemic**, because the native peoples had no immunity to the introduced diseases that had arrived on what Crosby termed "virgin soil."

The list of introduced diseases covers nearly everything known to modern medicine: smallpox, swine influenza, measles, tuberculosis, pertussis (whooping cough), anthrax, brucellosis, leptospirosis, and malaria. Some diseases were carried by the livestock that the Europeans brought with them, and others by the people themselves who had originally contracted them from their livestock. The most prolific hosts of disease were pigs, brought to the Americas as a food source on ships.

Genetic evidence suggests that early Americans had little immunity to disease because their ancestors arrived from remote regions in what is now Siberia—places that were isolated from the more disease-ridden cultures in warmer, wetter parts of southern Eurasia. Another reason for the lack of immunity among early Americans was the absence of livestock, which carry many diseases that can cross over to humans. The only livestock domesticated in the Americas were the llama and

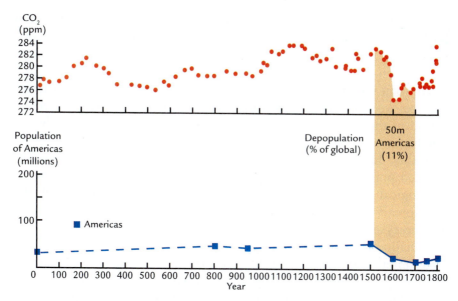

FIGURE 20-7 Changes in CO_2 concentration and the population of the Americas between 1400 and 1750. *[CO_2 record from C. McFarling Meure et al., "Law Dome CO_2, CH_4 and N_2O Ice Core Records Extended to 2000 Years BP,"* Geophysical Research Letters *33 (2006): L14810, doi:10.1029/2006GL026152. Drop in population based on W. M. Denevan, "The Pristine Myth: The Landscape of the Americas in 1492,"* Annals of the Association of American Geographers *82 (1992): 369–385.]*

alpaca in the mountains of South America, where cool temperatures tended to suppress disease.

Early Americans also lived in generally healthier environments, with better nutrition and little of the squalor typical of many European cities of the time (contaminated water sources, sewage running in gutters, human excrement thrown out of windows onto streets, and pervasive piles of horse manure). Arriving Europeans often remarked that early Americans were tall, impressive physical specimens, compared to the shorter, often pox-marked Europeans.

By 1500, many indigenous early American cultures had attained an impressive level of civilization (see Chapter 6). But between 1525 and 1750, these cultures all but vanished, leaving a small group of survivors, most of whom were so disorganized and demoralized that they soon forgot most of the achievements and skills of their predecessors.

Evidence for the scope of this unprecedented tragedy comes from several sources. In a few cases, direct eyewitness accounts are available from Europeans who entered the wilderness soon after initial contacts with native American peoples and there found abandoned villages, with human remains scattered by wild animals across the countryside, along with a few survivors still in shock at the loss of almost everyone they had known.

In some regions, especially Mexico, Peru, and the American southwest, archives kept by Spanish mission churches and small village centers chronicle the population collapses, although these records usually did not begin until somewhat after the initial and most lethal waves of mortality. In the 1960s, the anthropologist Henry F. Dobyns reconstructed population histories from these records. In addtion, recent archeological research has uncovered ever-increasing evidence of the widespread extent of large civilizations that existed before European conquest (see Chapter 6).

Smallpox was first recorded in the Caribbean by 1518, 26 years after initial contact and less than 10 years after the first permanent European occupation in the Americas. Coastal cultures were infected before inland ones because Europeans arrived by sea, and larger, wealthier civilizations were affected before smaller rural ones because Europeans were following rumors of gold.

The diseases moved inland ahead of the *conquistadores*, killing people who had never even seen a European. Pigs formed one disease vector, escaping into forested areas ahead of armies and passing diseases to deer and wild turkey, and on to the tribes that hunted them. In addition, natives who came down to coastal areas out of curiosity about the new arrivals contracted diseases and carried them back inland to their own tribes ahead of the Europeans.

Aztec Mexico

At the time of European arrival, 15 million people lived in the Aztec Empire, which extended from semi-arid highland forests in what is now central Mexico to lowland rain forests on the coast of the Gulf of Mexico. At the time, this empire was the most densely populated region on Earth. Over the preceding millennia, cultures in the region had developed a sophisticated agricultural system based on staples like maize, beans, squash, tomatoes, and peppers (see Chapter 6).

In 1519, Hernán Cortéz landed in Vera Cruz on the Gulf of Mexico coast, and within a year, smallpox outbreaks began in that region (Figure 20-8). Heading inland, with disease decimating people ahead of his small army, Cortéz was aided by alliances with other groups hostile to their Aztec rulers. He reached the capital Tenochtitlan (modern-day Mexico City), an amazing garden city built on a lake (see Chapter 6,

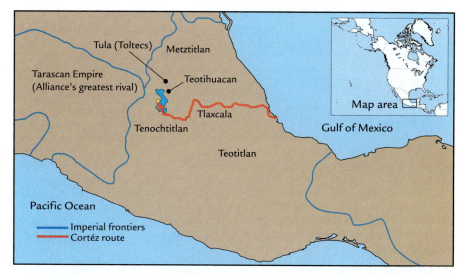

FIGURE 20-8 Hernán Cortéz's 1519–1520 march through the Aztec Empire to Tenochtitlan. *[Adapted from C. C. Mann, 1491(New York: Random House, 2006).]*

Figure 6-8). Meanwhile, smallpox continued to kill millions of Aztecs, including their leader Cuitláhuac, and the demoralized survivors began to feel that the gods were not on their side. The historians Sherborne Cook and Woodrow Borah estimated that only 25% of the original population was still alive by 1545, and just 3–4% by 1625.

Incan Peru

At its peak, the Inca Empire of some 14 million people extended almost 3,000 miles north and south along the Andes range in South America, and reached from the arid Pacific coast on the west to the warm rain forests east of the Andes. This empire was linked by a network of highways manned by runners who carried messages. Despite lacking metal tools, the Inca constructed buildings and fortresses made of monumental blocks of stone in cities like Qosqo ("Cuzco" in Spanish) and remote mountain sites like Machu Picchu.

Francisco Pizarro arrived at the Pacific coast of South America in 1531, but European diseases from other explorers had already gotten there first, with an outbreak of smallpox in 1526. As Pizarro and his brother Hernando moved southward through the empire (Figure 20-9), Cortéz's experiences in Mexico were repeated. European diseases went ahead of Pizarro's small army, killing countless people, including two successive heads of the Inca Empire, demoralizing the population and

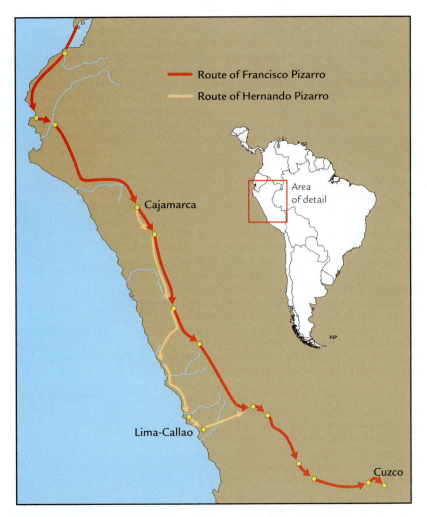

FIGURE 20-9 The marches of Francisco and Hernando Pizarro through the Inca Empire in 1531–1532. *[Adapted from G. Barraclough, ed.,* Concise Atlas of World History *(Maplewood, NJ: Hammond, 1982).]*

reducing resistance. Enemies of the Inca also joined forces with Pizarro. The confrontation culminated at the town of Cajamarca in November 1532 when Atahualpa, head of the empire, was taken prisoner and later killed. Major outbreaks of smallpox occurred in 1526, 1533, 1535, 1538, and 1565, along with outbreaks of influenza in 1538, typhus in 1546, diptheria in 1614, and measles in 1618. The anthropologist Henry Dobyns estimated a 90–95% mortality rate among the 14 million Inca.

Amazonia

The Amazon Basin held some 15 million people scattered over a vast area in villages and towns, most along the resource-rich east-central Amazon River with some farther out in the countryside (Figure 20-10). Food sources in the central Amazon rain forests included fruit orchards, manioc roots, and fish. People in the drier tree savanna of the Beni region far to the southwest redirected water from seasonal floods for crop irrigation (see Chapter 6).

The timing of the spread of disease across Amazonia is not well known because of the remoteness of many parts of that basin, much of which remains isolated even today. Villages and towns along the coast and in the lower reaches of the Amazon River were first struck by European diseases in the 1550s, while cultures in more remote inland areas probably succumbed later. As elsewhere, mortality rates of 85–90% left scattered and demoralized survivors.

North America

North America was less densely occupied than Central and South America, with some 7 million people. By the time of European arrival, maize and beans (originally domesticated in Mexico) had become major food sources, along with local crops such as sunflowers, and other food sources that are not in use today (see Chapter 6).

In 1539, Hernando de Soto landed in Tampa Bay, Florida with a privately funded army of 600 men, 200 horses, and 300 pigs. Over the

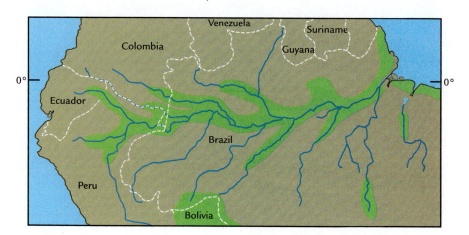

FIGURE 20-10 Early American occupation (*green shading*) of the Amazon Basin and Beni region. *[Adapted from C. C. Mann, 1491 (Random House, NY, 2006).]*

next four years, his expedition traveled across much of the southeastern and south-central interior of the present-day United States through the heartland of the Mississippian culture (Figure 20-11). He found most of the river valleys densely populated, with villages spaced every few kilometers or less and surrounded by earthen walls and moats (recall the Mississippian sites in Chapter 6, Figure 6-16). The intervening river valleys between the villages were filled with fields of maize and beans. Heading east into what is now north-central South Carolina, he came upon a region that had already been decimated by a disease outbreak caused by contacts a few years earlier with European traders along the Atlantic coast. Later in the voyage, his army crossed the Mississippi at present-day Memphis and moved west into what is now Arkansas, eastern Texas, and Oklahoma.

Hernando de Soto's pigs carried anthrax, brucellosis, leptospirea, and tuberculosis that ravaged the local populations. After another 150 years, with no further European contact, French trappers came down the Mississippi River in 1673 and found endless abandoned villages, some piled with the remains of bodies, overgrown with vegetation like cane and cypress. Even though de Soto only reached part of the area of the widespread Mississippian culture, trading along the rivers carried

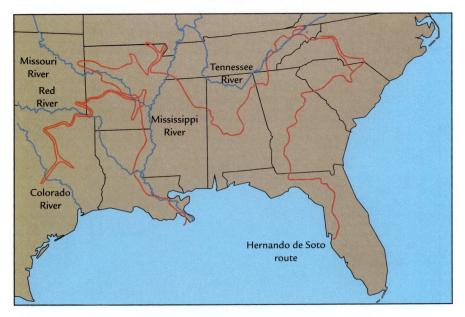

FIGURE 20-11 Hernando de Soto's 1539–1543 march into the interior of the present-day United States.

diseases farther into the interior to regions he had not explored. As the early Americans died, forests reclaimed river valleys that had been kept open by fire, as well as large areas of the upper Midwest that had been converted to prairie by frequent burning.

On the east coast of the present-day United States, native populations had been large enough prior to the 1600s to discourage Europeans from attempting to establish settlements (Figure 20-12). Later, as disease began to decimate the coastal populations, Europeans gained footholds in Massachusetts, Virginia, and elsewhere. As the European pandemic spread inland, large populations of indigenous people were swept away by diseases well before the Europeans actually appeared. By 1700, the native populations east of the Appalachian Mountains had been reduced to a tiny fraction of what they had once been. Many tribes like the Cherokee survived (for a while) because they lived in remote interior mountains.

FIGURE 20-12 Engraving of early American village of Secoton and maize fields in North Carolina by Theodor DeBry. *[The Art Gallery Collection/Alamy.]*

The demographic collapse in the Americas was arguably the greatest tragedy in all of human history. Not only did some 50 million people (85–90% of the previous population) die, but the survivors lost much of their knowledge of their own cultures and passed little if any of it along to their descendants.

In time, many historians and anthropologists came to doubt that these seemingly primitive cultures could have created the structures found through the Americas—stone buildings, monuments, roads, and massive earthen mounds. For a while, this evidence of higher civilizations was attributed to ancient peoples from Europe. Not until the twentieth century did the achievements of the indigenous peoples of the Americas begin to be seen for what they really were.

The American demographic collapse that began near 1525 coincides with the largest and most rapid CO_2 drop in the last 2,000 years (see Figure 20-7) and the largest CH_4 decrease in the last 1,000 years (see Figure 20-1). This episode appears to be the best candidate for exploring possible scientific links between population collapses and greenhouse-gas decreases.

Summary

Every interval of major decline in atmospheric CO_2 and CH_4 concentrations matches a time of massive population loss in Europe, China, or the Americas. This consistent relationship is highly suggestive, but the mechanisms that might have linked the population crashes and the greenhouse-gas declines need to be addressed.

Additional Resources

Borah, W. W., and Cook, S. F. *The Aboriginal Population of Central Mexico on the Eve of Spanish Conquest.* Berkeley, California: University of California Press, 1964.

Cartwright, F. F. *Disease and History.* New York: Dorsett Press, 1991.

Crosby, A. W. "The 'Virgin-Soil' Epidemic as a Factor in the Aboriginal Depopulation in America." *The William and Mary Quarterly* 33 (1976): 289–299.

Denevan, W. M., ed. *The Native Population of the Americas in 1492.* Madison: University of Wisconsin Press, 1978.

Denevan, W. M. "The Pristine Myth: The Landscape of the Americas in 1492." *Annals of the Association of American Geographers* 82 (1992): 369–385.

Diamond, J. *Guns, Germs, and Steel.* New York: W. W. Norton, 1999.

Dobyns, H. F. "Estimating Aboriginal American Population: An Appraisal of Techniques with a New Hemispheric Estimate." *Current Anthropology* 7 (1966): 395–416.

Mann, C. C. *1491*. New York: Random House, 2006.

McEvedy, C., and R. Jones. *Atlas of World Population History*. New York: Penguin, 1978.

McNeill, W. H. *Plagues and Peoples*. New York: Anchor, 1998.

Shaffer, L. N. *Native Americans before 1492: The Moundbuilding Centers of the Eastern Woodlands*. Armonk, NY: M. E. Sharpe, 1992.

CHAPTER 21

EFFECTS OF HUMANS ON SHORT-TERM GREENHOUSE-GAS REDUCTIONS

Chapter 19 showed that natural factors do not explain most of the amplitude of prominent short-term drops in CO_2 and CH_4 during the last two millennia, and Chapter 20 revealed a highly suggestive correlation between these drops and intervals of massive human mortality (Figure 21-1). But what mechanisms could link population decreases to reductions in greenhouse-gas concentrations?

Anthropogenic Role in Short-Term CO$_2$ Decreases

Population decreases are likely to have contributed to short-term drops in atmospheric CO_2 during the last 2,000 years primarily through the process of reforestation. Episodes of mass human mortality caused either by pandemics or civil strife would have resulted in abandonment of land formerly used for agriculture, which would have allowed regrowth of forests on the abandoned land. This return of the forests would have stored carbon from atmospheric CO_2 in the wood of the growing trees.

Mechanisms

This proposed link between mass mortality and reforestation can be evaluated in the context of the number and density of the human populations

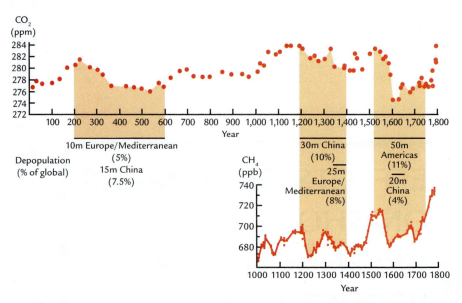

FIGURE 21-1 CO_2 and CH_4 decreases during the last 2,000 years correlate with intervals of massive population loss. [CO_2 record from C. McFarling Meure et al., "Law Dome CO_2, CH_4 and N_2O Ice Core Records Extended to 2000 Years BP," Geophysical Research Letters 33 (2006): L14810, doi:10.1029/2006GL026152. Population levels from C. McEvedy and R. Jones, Atlas of World Population History (New York: Penguin Books, 1978) and from W. M. Denevan, "The Pristine Myth: the Landscape of the Americas in 1492," Annals of the Association of American Geographers 82 (1992): 369–385. Methane concentrations from L. E. Mitchell, E. J. Brook, T. Sowers, J. R. McConnell, and K. Taylor, "Multidecadal Variability of Atmospheric Methane, 1000–1800 C.E.," Journal of Geophysical Research 116 (2011), GO2007, doi:10.1029/2010JG001441.]

living on the land in the regions where calamities occurred. The effects of pandemics and civil strife would have been far less severe in places where populations had grown very dense, most of the arable land was already in cultivation, and a significant fraction of the population had moved to towns and cities. Under these conditions, abandoned farms could have been quickly reoccupied by family members or by neighbors who seized an opportunity to claim and farm desirable plots of land. As a result, net farm abandonment (and carbon storage in growing forests) in these regions would likely have been much smaller than implied by the total mortality numbers. In contrast, for less densely populated regions that suffered extremely high rates of mortality, the few survivors would not have been able to replace those who had died. In these cases, most of the once-farmed land would have gradually reverted to forest.

Given these different possibilities, the amount of reforestation caused by each episode of mass mortality and the resulting effect of subsequent carbon storage on atmospheric CO_2 must be evaluated on a case-by-case basis. Three intervals in the last 2,000 years that show decreases in both population and CO_2 merit attention: (1) the interval from 200 to 600, (2) the medieval interval that began at 1200 and continued erratically to 1400; and (3) the interval from the early 1500s through the early 1600s (see Figure 21-1).

The CO_2 decrease from 200 to 600 was modest in size (5 ppm) despite the fact that millions of people died (about 10 million in Europe and about 15 million in China). Because this interval lasted almost 400 years, ongoing deforestation elsewhere on Earth would have offset part of the effect of gradual reforestation in the stricken regions. In addition, moderately high population densities already present in the Roman Empire and China may have limited the amount of land actually abandoned during this time because many people were available to reoccupy much of the land.

Similarly, the even higher population densities reached in China by the time of the civil strife in the 1200s and in Europe by the Black Death pandemic from 1348 to 1353 probably limited land abandonment and reforestation, even though tens of millions of people died in each case. The rapid population rebound in Europe in the century and a half following the Black Death (and consequent reoccupation of many farms) would also have limited the net amount of reforestation.

The virgin soil pandemic in the Americas during the 1500s and 1600s was an entirely different matter. Not only was the total mortality enormous (about 50 million people), but the 85–90% mortality rate also left few survivors to reoccupy and farm abandoned land. Waves of mass mortality swept into the Aztec Empire of Central America with Cortez, the Incan Empire of South America with Pizarro, and southeastern North America with de Soto, all during the early to middle 1500s. In these regions, population levels then remained depressed for centuries as new outbreaks of disease continued to decimate survivors. This unprecedented population collapse in the Americas was not reversed until some 250 years later, when Europeans began to arrive in sufficient numbers to clear large areas of forest during the middle and late 1700s. As a result, most areas that were reforested in the 1500s and 1600s remained that way for 200 to 250 years.

Quantifying Human Impacts on Short-Term CO_2 Decreases

Transforming the amount of reforestation in calamity-stricken areas into estimates of net carbon storage in newly growing forests requires several assumptions, each beset by sizeable uncertainties. The amount of carbon stored in the regrowing forests is the product of three variables: the number of people who died and left their farmed land abandoned; the average per capita amount of land these farmers had previously kept clear of natural vegetation; and the net gain of carbon stored per unit area in the regrowing forests compared to the amount of carbon that had existed in the crops or pastures that were tended prior to the calamity. The calculation can be written out as follows:

(number of people) $\times$ (area farmed per person) $\times$ (tons of carbon stored per area) = tons of carbon

The total carbon storage estimated in this way has to be weighed against the ongoing deforestation elsewhere in the world that would have continued to add CO_2 to the atmosphere. These carbon emissions in regions not affected by the calamity would have countered part of the carbon storage occurring in the stricken areas.

The few published attempts at estimating net carbon storage during the three intervals of interest yield widely different results, from values so small they had only negligible effects on atmospheric CO_2 concentrations (in one estimate, a decrease of less than 1 ppm) to numbers large enough to explain much of the observed CO_2 decreases. Variations among these estimates arise mainly from differing choices of values for the factors listed above: mortality, per capita land use, and carbon density in the regrowing forests.

Jed Kaplan and colleagues were the first scientists to use historical data to quantify the per capita land use of cultures in different stages of agricultural development and population density (see Chapter 8). Their simulation shows that the long-term increase in cumulative carbon emissions because of deforestation was interrupted three times during the last 2,000 years by pauses or reductions in cumulative carbon emissions caused by net carbon storage in regrowing forests (Figure 21-2). The first two intervals marked by small amounts of carbon storage coincide with times of historically defined population reductions: late Antiquity (200–600) and the late Medieval Era (1200–1400). The third interval, coinciding with the American population collapse, is marked by a large carbon storage of about 37 billion tons.

Translating the 37 billion tons of estimated carbon storage into a net reduction in atmospheric CO_2 concentrations introduces several additional complications. One problem is that precise (decade-by-decade) time histories of human mortality and reforestation are not well defined for the period from the early 1500s through the 1600s. A second complication is

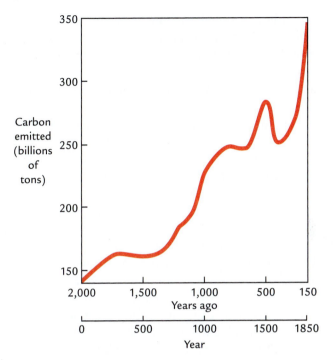

FIGURE 21-2 Cumulative emissions of carbon during the last 2,000 years. *[Adapted from J. O. Kaplan, K. M. Krumhardt, E. C. Ellis, W. F. Ruddiman, C. Lemmen, and K. K. Goldewijk, "Holocene Carbon Emissions as a Result of Anthropogenic Land Cover Change," The Holocene 25 (2011): 775–791, doi:10.1177/0959683610386983.]*

that the rate of carbon storage in forest wood lagged behind the time of highest mortality because of the many decades required for growing forests to store carbon. In most regions, the new growth would have started out as shrubs and small saplings, then grown into mid-sized trees, and eventually become large mature forests. Also, in regions of tropical rain forest, bamboo was grown near many settlements for use in constructing dwellings and fence posts. Within 10 or 20 years of abandonment, dense stands of bamboo would have expanded across previously cleared land and stored a substantial amount of carbon. Consistent with Kaplan's estimate of decreasing carbon emissions in Figure 21-2, the largest rates of carbon storage should have occurred in the middle or late 1500s to the early 1600s, when forests were still growing across large areas hit hardest by the early arrival of disease in the 1520s and 1530s.

   According to Kaplan, the large rates of rapid reforestation of the Americas beginning in the mid-1500s were more than sufficient to off- set the slower deforestation occurring in other regions. His simulation shows that cumulative global carbon emissions did not return to the levels that had been reached in the 1400s until Europeans began clearing large amounts of American forests in the late 1700s (see Figure 21-2). Even today, many areas in the interior of the Amazon Basin that had been cleared prior to 1500 remain almost completely unoccupied (see Chapter 6).

The next step, after calculating the effect of reforestation on carbon storage, is to connect this carbon storage history to changes in atmospheric CO_2 concentration. This step is complicated because the atmosphere was not the only source of carbon for reforestation. Other carbon reservoirs would also have contributed, thereby reducing the net draw on the atmosphere. For example, today's human activities are emitting an excess of fossil-fuel carbon into the atmosphere (Figure 21-3A). On a short-term (yearly) basis, the atmosphere takes up a little more than half (55%) of the released CO_2, another 25–30% is stored in the ocean, and the rest ends up in global vegetation, mainly because of faster plant growth fertilized by rising atmospheric CO_2 levels. Our present rate of injection of CO_2 into these reservoirs is so rapid that it overwhelms other slower exchange processes that will become more important in the future. Over centuries to millennia, the deep ocean will play the dominant role by absorbing most of the pulse of fossil-fuel CO_2 we have emitted (see Chapter 17).

A similar process occurred during the American pandemic and population collapse, but in reverse (Figure 21-3B). Initially, the CO_2 needed for reforestation had to be taken from the atmosphere, but almost immediately the shallow ocean and the vegetation in other regions would have begun to contribute some of their own carbon, reducing the net draw on the atmosphere.

If all of the CO_2 had come from the atmosphere, Kaplan's estimate of 37 billion tons of carbon storage would have reduced the atmospheric concentration by about 17 ppm, more than the observed drop of 10 ppm (see Figure 21-1). However, if the shallow ocean and vegetation joined in within a few years and contributed 50% of the total carbon, the draw on atmospheric CO_2 would have been reduced by half, to about 8.5 ppm. This value falls in the middle of the observed 7–10 ppm drop.

Over time, the deep ocean would have begun to play a larger role. If the net CO_2 burial of 37 billion tons had persisted for 1,000 years or more, the deep ocean would have had enough time to contribute its full share of carbon, and the drawdown of atmospheric CO_2 would have been reduced to just 2.6 ppm. But well before that could happen, the wave of new deforestation and CO_2 emissions by European immigrants had begun.

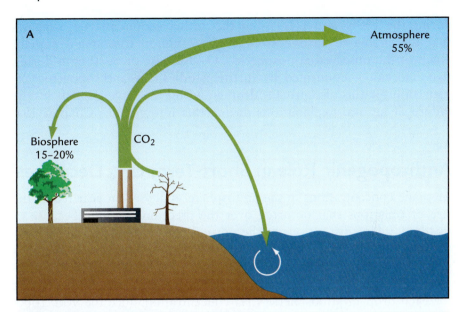

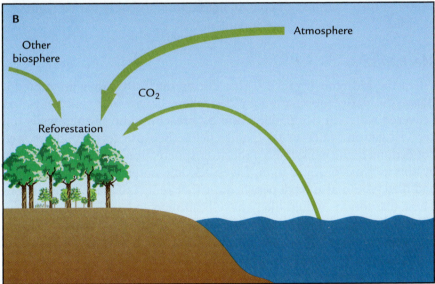

FIGURE 21-3 (A) Present-day emissions of carbon end up in the atmosphere (about 55%), the ocean (roughly 25–30%), and in vegetation (about 15–20%). (B) Carbon stored in American forests after 1525 would have been drawn from these same reservoirs in roughly the same proportions as in A.

More sophisticated analyses of the quantitative effect of American reforestation on the atmospheric CO_2 decrease will require accurate information on both the amount and the timing of carbon storage during the 1500s and 1600s. But this rather simple analysis suggests that the estimated carbon storage of 37 billion tons might explain much of the observed atmospheric CO_2 decrease of 7–10 ppm during that interval. If confirmed by further studies, the anthropogenic explanation for this abrupt CO_2 drop could fill the large gap left by the failure of natural explanations during this interval (see Chapter 19).

Anthropogenic Role in Short-Term CH_4 Decreases

Methane concentrations in the atmosphere in any given year reflect the size of emissions that occurred over the preceding decade. Because of this link, major disruptions to the normal rate of emissions in the past would have altered the CH_4 concentration in the atmosphere within a period of a few years to at most a decade. The historical records summarized in Chapter 20 suggest plausible mechanisms that could account for each drop in atmospheric CH_4 concentration since the year 1000, although the exact mix of controlling factors probably varied from case to case.

Mechanisms

When the Mongols invaded China during the 1200s, they attempted to starve the native population by destroying canals and sluiceways that delivered water to rice paddies and by breaching the earthen walls that kept irrigation water in the paddy fields. Because fewer paddies were flooded, methane emissions to the atmosphere would have been reduced by this pervasive destruction. Livestock left untended by farmers who had died would also have perished, further reducing methane emissions. These disruptions would likely have been repeated in the social upheavals in China during the 1600s. In both cases, these disruptions to the normal operation of society are plausible contributors to the measured drops in atmospheric CH_4.

Because there were no rice paddies in medieval Europe, the Black Death (during the middle 1300s) may not have been a large factor in long-term methane emissions, although losses of European livestock could have played some role. In the case of the American demographic collapse that began near 1525, rice paddies again could not have been a factor, and relatively few livestock were tended (only llama and alpaca in the Andes). But early Americans had previously burned extensive areas to keep grasslands open, so as 85–90% of the population across the Americas perished (see Chapter 20), the biomass burning that they had previously carried out in oxygen-deficient environments would have diminished, along with methane emissions.

Quantifying Human Impacts on Short-Term CH₄ Decreases

One method to evaluate the combined effect of the mechanisms proposed above is to compare the relative size of the decreases in global population during each interval against estimates of the decreases in anthropogenically generated CH_4 in the atmosphere. According to the early anthropogenic hypothesis, between the years 1200 and 1600 the anthropogenic CH_4 anomaly added an estimated 200–250 ppb to the natural trend (see Chapter 10, Figure 10-5). This value is derived by calculating the difference between the measured CH_4 increase during that interval and a straight-line extrapolation of the natural decrease toward the proposed 450-ppb level. If this estimate is correct, then the short-term CH_4 decreases of 15–25 ppb that occurred between 1200 and 1600 (see Figure 21-1) represent 6–12% of the total anthropogenic contribution at that time.

In comparison, global population during that interval increased from about 400 million people in the year 1200 to about 600 million people by the year 1600 (see Part 2, Figure 2i-2). The 55 and 70 million people that died during the two major mortality episodes shown in Figure 21-1 represented 15–18% of the total global population at the time. Based on this somewhat crude comparison, the large proportional decreases in population (15–18%) appear to be more than sufficient to explain the smaller proportional drops (6–12%) in estimated anthropogenic CH_4 concentrations.

Other Anthropogenic Impacts: Albedo Changes Due to Forest Clearance

Another anthropogenic factor that contributed to regional Little Ice Age cooling late in pre-industrial times (1400–1800) was the change in surface albedo caused by cutting high-latitude forests. Clearance of evergreen trees at high latitudes results in cooling because snow-covered pastures and croplands reflect more solar radiation back to space than the dark treetop canopies of evergreen spruce and fir forests (see Chapter 16, Figure 16-2).

Most of the lower mid-latitude forests of southern, western, and central Europe had been removed long before the time of the Little Ice Age, but those at higher latitudes of northeastern Europe from Poland to western Russia had remained largely intact. By 1400, the lumber needs of rapidly growing western European economies drove widespread deforestation in those regions (see Chapter 4), and by 1800, about 70% of those forests had been cut (Figure 21-4). Cutting in these areas would have increased the reflectivity of snow-covered land surfaces and cooled climate just east of the regions that provide some of the strongest evidence of Little Ice Age cooling. In east-central North America, forests had also begun to be heavily logged starting around 1750. The quantitative effect of late pre-industrial deforestation in both of these regions needs further investigation.

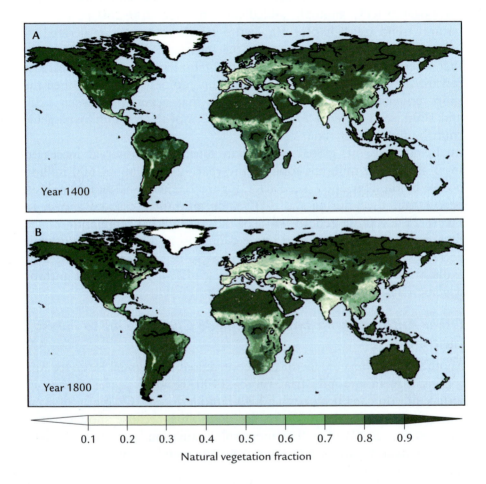

FIGURE 21-4 Estimates of the percentage of natural vegetation cleared by anthropogenic activity as of the years 1400 (A) and 1800 (B). *[Adapted from J. O. Kaplan, K. M. Krumhardt, E. C. Ellis, W. F. Ruddiman, C. Lemmen, and K. K. Goldewijk, "Holocene Carbon Emissions as a Result of Anthropogenic Land Cover Change,"* The Holocene *25 (2011): 775–791, doi:10.1177/0959683610386983.]*

Effect of Anthropogenic CO_2 and CH_4 Decreases on Climate

Did anthropogenically driven drops in greenhouse-gas concentrations play a significant role in the general cooling into the Little Ice Age? For the short-term 15–25 ppb drops in methane between 1000 and 1700 (see Figure 21-1), the answer is simple: they did not. Standard estimates of global climate sensitivity to CH_4 changes of that size suggest net temperature decreases of only 0.01°C, a trivial fraction of the reconstructed temperature variations of the Little Ice Age (Figure 21-5).

The largest change in atmospheric CO_2 concentration during the same interval is the 10-ppm drop during the 1500s into the early 1600s. Estimates of global climate sensitivity suggest that this abrupt CO_2 decrease could have caused a net cooling of about 0.13°C. If the (mainly anthropogenic) reduction in atmospheric CO_2 between the 1500s and 1600s accounted for a 0.13°C cooling, and if the average estimated (northern hemisphere) cooling at this time was 0.17°C (see Chapter 19), humans appear to have been responsible for much of the cooling during this brief interval.

As noted earlier, however, proxy-indicator reconstructions of northern hemisphere climate present varying pictures of temperature change during this interval and across much of the rest of the millennium-long

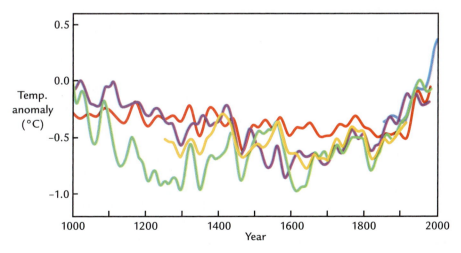

FIGURE 21-5 Several reconstructions of changes in northern hemisphere temperature during the last 1,000 years. *[Adapted from P. D. Jones and M. E. Mann, "Climate Over Past Millennia,"* Reviews of Geophysics 42 *(2004), No. 2, RG2002, doi:10.1029/2003RG000143.]*

record (see Figure 21-5). In addition, reliable temperature reconstructions for the Southern Hemisphere are not yet available. Until scientists can come to closer agreement about the shape of the global temperature trend during the last millennium, it will be difficult to assess the role of anthropogenic CO_2 variations in these temperature changes.

Summary

Plausible mechanisms link short-term drops in CO_2 and CH_4 to massive decreases in population during the last two millennia. Reforestation of abandoned cropland pulled CO_2 out of the atmosphere and stored it in the growing vegetation, while disruption of rice irrigation infrastructure and livestock deaths reduced CH_4 emissions to the atmosphere. The methane decreases did not lead to significant temperature drops, but the CO_2 decrease from the middle 1500s to early 1600s could have contributed to a climatic cooling that formed part of the longer-term trend into the Little Ice Age.

Additional Resources

Faust, F. X., C. Gnecco, H. Mannstein, and J. Stamm. "Evidence for the Post-conquest Demographic Collapse of the Americas in Historical CO_2 Levels." *Earth Interactions* 10 (2006): 1–14. doi:10.1175/EI157.1.

Ferretti, D. F., et al. "Unexpected Changes to the Global Methane Budget over the Last 2000 Years." *Science* 309 (2005): 1714–1717.

Kaplan, J. O., K. M. Krumhardt, E. C. Ellis, W. F. Ruddiman, C. Lemmen, and K. K. Goldewijk. "Holocene Carbon Emissions as a Result of Anthropogenic Land Cover Change." *The Holocene* 25 (2010): 775–791. doi:10.1177/0959683610386983.

Kauffman, D. S., et al. "Recent Warming Reverses Long-Term Arctic Cooling." *Science* 325 (2009): 1236–1239. doi:10.1126/science.1173983.

Nevle, R. J., D. K. Bird, W. F. Ruddiman, and R. A. Dull. "Neotropical Human-Landscape Interactions, Fire, and Atmospheric CO_2 during European Conquest." *The Holocene* 25 (2011): 853–864. doi:10.1177/0959683611404578.

Part 6 Summary

Natural (orbital) decreases in summer solar insolation at high northern latitudes have been slowly pushing climate toward a cooler state for several millennia (Figure 6s-1). Parts 4 and 5 of this book proposed a new paradigm in which permanent snow cover and small ice sheets would have begun growing at high northern latitudes in the last few thousand years, had it not been for the warming caused by greenhouse-gas emissions to the atmosphere from early farming. Over brief (decade-long) intervals, volcanic eruptions added to this general cooling trend.

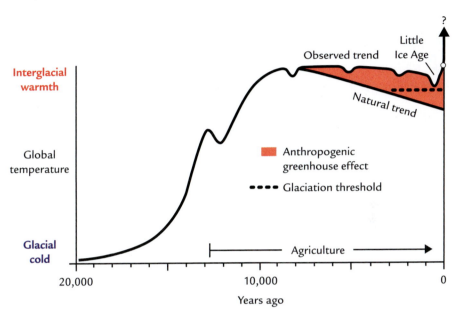

FIGURE 6s-1 Decreasing summer solar radiation has driven a gradual long-term cooling for thousands of years in high northern latitudes, although anthropogenic greenhouse gases have offset much of the cooling and kept climate relatively warm. For decade-long or century-long intervals, volcanic eruptions and massive population calamities have added to the cooling.

Several abrupt decreases in CO_2 and CH_4 concentrations during the last 2,000 years cannot be explained by natural factors and must have largely resulted from massive population losses caused by pandemics and civil strife. The largest calamity (the arrival of European diseases in the Americas) killed some 50 million people between the early 1500s and early 1600s and drove atmospheric CO_2 and CH_4 levels down. The cooler Little Ice Age interval came to an end during the mid-1800s, when rising greenhouse-gas concentrations produced mainly by clearing forests and burning fossil fuels caused climate to warm.

Could future pandemics, perhaps global in extent, boost carbon storage in regrowing forests enough to once again drive down atmospheric CO_2 concentrations? This scenario seems highly unlikely. Agriculture in many countries is now heavily mechanized, and very few farmers are needed to drive the machines that cultivate large expanses of land. Even if many hundreds of millions of people died, more than enough people would still be left to drive the plows, tractors, and combines needed to grow and harvest crops.

For most of the tens of thousands of years since humans evolved on this planet, we lived by foraging for food. Our ancestors collected seeds from wild grasses, dug roots and tubers, ate fruit and nuts that grew wild, and hunted animals and caught fish, but human populations never exceeded a few million people. For millions of years before that, our human-like predecessors had subsisted in much the same way and almost certainly at even lower population levels. Populations never gained much ground during all that time because of the constraints imposed by the uncertainties and high mobility of a foraging and hunting/gathering existence.

Starting near 10,000 years ago, these constraints began to give way to a revolutionary new way of subsisting: staying mostly in one place, growing crops, and herding livestock. Because of variations in the kinds of wild plants and animals that were suitable for domestication in different regions, farming took on many different forms as it appeared across a span of several thousand years. But in region after region, the success of agriculture in feeding people triggered an explosion of human populations, and farmers began to spread widely across Earth's arable regions after 8,000 years ago, displacing or outbreeding hunter-gatherers. Within a few thousand years, most people on Earth had become farmers, and in time urban civilizations began to appear in many areas.

Meanwhile, a second, somewhat enigmatic, change was also underway. Starting near 7,000 years ago, atmospheric carbon dioxide (CO_2) concentrations (as measured in ice-core air bubbles) began to rise, followed by a similar increase in methane (CH_4) beginning 5,000 years ago. Both of these greenhouse-gas increases appear enigmatic because they were opposite in direction to the downward gas trends during equivalent intervals of previous interglaciations (Figure E-1).

These CO_2 and CH_4 increases are particularly intriguing because they coincided with the time that farming was spreading across Earth. This similarity in timing raises the possibility that human agriculture played a role in boosting the atmospheric concentrations of these two greenhouse gases. This possibility has a reasonable scientific basis. Clearing forests to let in the sunlight needed to grow crops or create new pastures emitted extra CO_2 into the atmosphere. Irrigating rice paddies, tending herds of livestock, and burning grasslands and crop residues emitted CH_4 beyond the amounts from natural sources. The spread of farming activities over many thousands of years is obviously

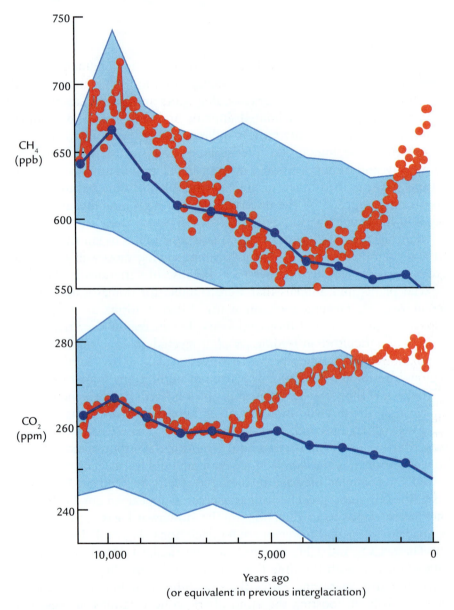

FIGURE E-1 CH₄ trend (top) and CO₂ trend (bottom) during the current interglaciation (*red*) compared to the average (*dark blue*) and standard deviation (*light blue*) of previous interglaciations. [*CH₄ and CO₂ values from EPICA Community Members, "Eight Glacial Cycles from an Antarctic Ice Core,"* Nature *429 (2004): 623–628.*]

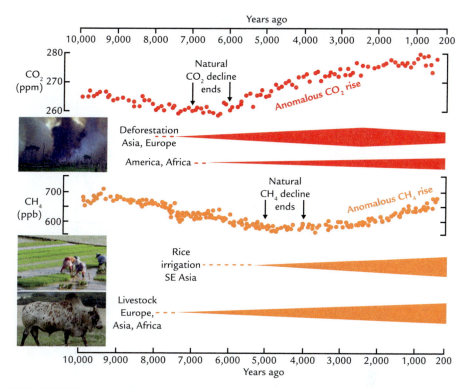

FIGURE E-2 Coincident timing of anomalous long-term rises in atmospheric CO_2 and CH_4 concentrations and spread of gas-emitting agricultural activities across the continents. *[CO_2 and CH_4 values from EPICA Community Members, "Eight Glacial Cycles from an Antarctic Ice Core," Nature 429 (2004): 623–628. Photos: (top) Frits Meyst/Hollandse Hoogte/Redux; (middle) dbimages/Alamy; (bottom) blickwinkel/Alamy.]*

a factor to consider in the anomalous rises in the CO_2 and CH_4 concentrations (Figure E-2).

No scientist aware of this evidence questions the notion that early agriculture had *some* effect on greenhouse-gas concentrations during those early (pre-industrial) millennia, but the size of the effect is uncertain. And so, for more than a decade, the scientific community has been actively debating the amount and timing of agricultural clearance and its effects on greenhouse-gas emissions. Were those effects trivially small, or did they contribute significantly to the observed CO_2 and CH_4 increases during the last several thousand years? During the past decade, some forty papers have been published favoring one side or the other of this debate, and every year news media have published several articles covering this issue.

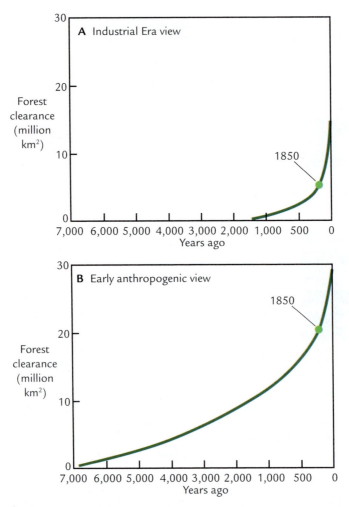

FIGURE E-3 Two highly schematic views of the history of cumulative forest clearance: (A) Industrial Era view; (B) early anthropogenic view. (Note the scale change at 1,000 years ago.)

According to a long-standing "Industrial Era view" (Figure E-3A), pre-industrial land clearance by farming was small, and the resulting effects on atmospheric CO_2 and CH_4 concentrations were trivial compared to the much larger changes during the Industrial Era (since 1850). In contrast, a newer (and directly opposing) "early anthropogenic view" holds that pre-industrial clearance for agriculture transformed Earth's surface enough to account for the entire pre-industrial increases in CO_2 and CH_4 concentrations (Figure E-3B), while also offsetting what would have been natural decreases in the concentrations of both gases (see Figure E-1).

Models that simulate past land use and resulting carbon emissions from forest clearance have been employed in this debate. First-generation simulations that assumed a nearly one-for-one relationship between clearance and past population levels predicted very little pre-industrial clearance and a small CO_2 increase of just a few ppm, because populations during this time period were only a few percent of those today. These results supported the Industrial Era view and contradicted the early anthropogenic one.

But those simulations ignored a range of historical evidence showing that farmers in both China and Europe used much more land per person 2,000 years ago than they did in later pre-industrial centuries. As ever-increasing populations shrank the amount of available land, farmers gradually learned to extract more food from each hectare.

Newer model simulations that incorporate this historical evidence indicate far greater pre-industrial clearance and much larger CO_2 emissions. In more heavily populated regions such as Europe, China, and India, these newer simulations reveal nearly complete deforestation well before the Industrial Era, in contrast to the persistence of nearly pristine forests into recent centuries simulated by the earlier models. Evidence from fieldwork in several disciplines supports these historically based reconstructions and refutes the simulations showing very little clearance in late pre-industrial times.

Archeological evidence has also been brought to bear on this debate. Anthropogenic CH_4 emissions have been estimated based on the spread of irrigated rice paddies across southern Asia between 5,000 and 1,000 years ago. The levels of methane emissions simulated from this evidence account for most of the measured rise of atmospheric CH_4 during that interval. In addition, the simultaneous spread of methane-emitting livestock revealed by archeological data from Asia, Africa, and Europe during the last several thousand years, along with increased biomass burning on all continents, must have added to total methane emissions, although neither of these contributions has yet been quantified.

The debate continues over the cause of the gradual CO_2 and CH_4 increases during the last 7,000 years, but the emerging historical and archeological data favor a large early anthropogenic role in pre-industrial times. If the early anthropogenic view continues to gain ground because of this kind of evidence, its corollary predictions will also merit greater attention.

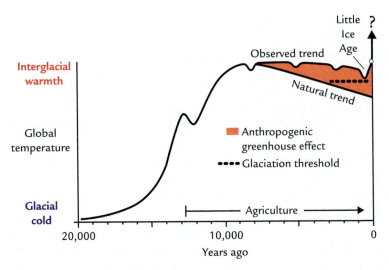

FIGURE E-4 Hypothesized effect of pre-industrial greenhouse-gas emissions in keeping global temperature warm enough to avoid a glaciation that would have begun under natural conditions.

One such prediction involves the effect of agriculture on climate. Natural climatic trends during the last 10,000 years have been gradually pushing climate toward cooler conditions in polar and subpolar regions. But according to the early anthropogenic view, these regions would have become even cooler by now, had it not been for the greenhouse-gas emissions from agriculture (Figure E-4). Without agriculture, CO_2 concentration in the atmosphere would have fallen to an estimated 240–245 ppm, instead of rising to the observed 280–285 ppm. The CH_4 level would have fallen to 445–450 ppb, rather than rising to 790 ppb. If gas concentrations had fallen this far, one predicted result would have been the onset of small-scale glaciations.

Climate model simulations that omit both the Industrial Era greenhouse gases (measured in ice cores) and the pre-industrial greenhouse gases (proposed in the early anthropogenic hypothesis) show year-round snow cover persisting in parts of northern Canada and Eurasia (Figure E-5). Snow cover that persists through warm summer months is an indication of glacial inception, because the snow will then grow deeper in the cold months (and future years) that follow. If the reduced greenhouse-gas levels assumed for these simulations are valid, the results justify the provocative claim that early farming stopped a new glaciation from taking hold in these far-northern regions sometime

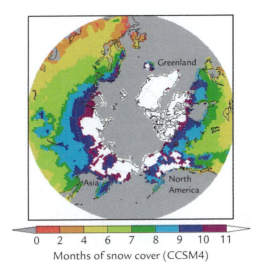

0 2 4 6 7 8 9 10 11
Months of snow cover (CCSM4)

FIGURE E-5 Simulated months of summer snow cover for an experiment with the CCSM4 model using the lower CO_2 and CH_4 values proposed in the early anthropogenic hypothesis. *[From work in progress by S. Vavrus, F. He, and J. Kutzbach.]*

before the Industrial Era. Now, the extra warmth from the additional greenhouse gases released since 1850 has made a new northern hemisphere glaciation impossible, putting a definitive end to glacial cycles that had persisted for almost 3 million years.

Another prediction of the early anthropogenic hypothesis is the claim that massive decreases in human populations caused by major calamities (such as pandemics and civil strife) caused short-term drops in CO_2 and CH_4 concentrations during historical times (Figure E-6). As populations dropped, cultivated lands reverted to forest, and atmospheric CO_2 was stored (through photosynthesis) in the wood of the growing trees. The largest calamity was the depopulation of some 50 million early Americans during the 1500s (80–90% of the population) because of their lack of natural immunity to diseases introduced by Europeans.

Population drops in Asia occurred mainly because of civil strife, such as Genghis Khan's conquest of China during the 1200s. Methane emissions were reduced because Khan raiders destroyed the infrastructure that delivered water to irrigate rice paddies. As people starved and perished by the millions, livestock also died, further reducing methane emissions.

The ongoing argument over natural versus anthropogenic influence on land use, greenhouse gases, and climate during recent millennia sets

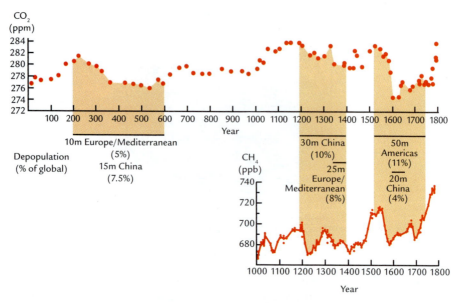

FIGURE E-6 CO_2 and CH_4 decreases during the last 2,000 years correlate with intervals of massive population loss. [*CO_2 record from C. McFarling Meure et al., "Law Dome CO_2, CH_4 and N_2O Ice Core Records Extended to 2000 Years BP," Geophysical Research Letters 33 (2006): L14810, doi:10.1029/2006GL026152. Population levels from C. McEvedy and R. Jones, Atlas of World Population History (New York: Penguin Books, 1978); and from W. M. Denevan, "The Pristine Myth: The Landscape of the Americas in 1492," Annals of the Association of American Geographers 82 (1992): 369–385. Methane concentrations from L. E. Mitchell, E. J. Brook, T. Sowers, J. R. McConnell, K. Taylor, "Multidecadal Variability of Atmospheric Methane, 1000–1800 C.E.," Journal of Geophysical Research 116 (2011), GO2007, doi:10.1029/2010JG001441.]*

forth two strikingly different views of the prehistory and history of our species. Either human civilization developed over thousands of years on a planet favored by the warmth of a natural interglaciation, or, as I believe, we played a significant role in producing some of the interglacial warmth within which our civilizations came into being.

GLOSSARY

aerosols Extremely small particles or droplets of natural or human origin carried in suspension in the air.

albedo The decimal fraction or percentage of incoming solar radiation reflected from a surface.

albedo-temperature feedback A positive feedback that amplifies an initial temperature change by altering the amount of snow cover or sea ice and changing the amount of solar radiation absorbed by Earth's surface.

anthromes Areas transformed in varying degrees from their natural state by humans.

Anthropocene The interval of dominant human influences on climate and the environment.

anthropogenic Caused by humans (usually referring to changes in climate, ecosystems, or other parts of the environment).

ard plow (also *scratch plow*) A primitive wooden plow used to incise "x" patterns in grassy soils to make them easier to turn with a shovel.

biomes Ecosystems that occupy broad areas in response to varying patterns of temperature and precipitation (often referring to vegetation units).

biosphere The global Earth system that supports life; it includes the oceans, land surfaces, soils, and atmosphere.

Black Death A massive outbreak of bubonic plague that killed one of every three people in Europe between 1348 and 1352.

Bronze Age The interval between 4,500 and 2,500 years ago when many cultures made tools of bronze, following the earlier interval of stone tool use and preceding the later invention of iron tools.

brown clouds Layers of particulate debris, floating at an elevation of 2–3 km in the atmosphere, which are produced by cook stoves and other fuel combustion. These clouds absorb incoming solar radiation and cool the surface.

carbon/climate model A numerical simulation of Earth's physically based climatic processes and the movement of carbon through its systems.

carbon dioxide (CO_2) A greenhouse gas formed by the oxidation of carbon, both through natural processes such as fires and rotting of vegetation and through anthropogenic processes such as burning of fossil fuels and forests.

carbon isotopes Forms of the element carbon with different atomic masses. Used to trace the movement of different kinds of carbon (^{12}C and ^{13}C) through Earth's climate system and to measure elapsed time indicated by radioactive decay (^{14}C and ^{12}C).

carbonate compensation hypothesis The idea that a delayed adjustment of ocean chemistry to changes that occurred late in the preceding deglaciation explains the atmospheric CO_2 increase during the last several thousand years.

climate Fluctuations in Earth's air, water, ice, vegetation, and other properties on time scales longer than one year, in contrast to shorter-term changes over days to months, which are called *weather*.

climate engineering Intentional manipulation of climate by humans.

climate proxy A quantifiable indicator of climate change recorded in natural systems. Examples include tree rings, ice cores, and corals.

CO_2 fertilization The increased growth rate of plants caused by adding carbon dioxide to the atmosphere.

CO_2 solubility The amount of carbon dioxide dissolved in seawater, which decreases with increasing temperature.

COHMAP (Cooperative Holocene Mapping Project) A collaborative research effort in the 1980s and 1990s that evaluated the causes of climate change during the most recent deglaciation by comparing geologic data with climate model simulations.

conifer forest Forests of evergreen, needle-bearing trees.

coppicing Pruning trees to promote growth of many small branches that can be easily cut to make fencing, and for other agricultural uses.

coral reef hypothesis The idea that the increase in atmospheric CO_2 during the last several thousand years resulted from increased rates of construction of coral reefs.

deforestation Cutting and burning of forests by humans to clear land for agriculture and other uses.

dew point The temperature at which cooling air becomes fully saturated with water vapor and condensation to liquid form can occur.

dibbles Digging sticks for planting seeds.

Domesday Book A census of population and land use in Britain compiled for William the Conqueror, published in 1089.

domestication A gradual process in which wild plants and animals are transformed by artificial selection into forms that are better suited for human agriculture.

early anthropogenic hypothesis The idea that humans began interfering with atmospheric greenhouse-gas concentrations and climate thousands of years ago.

eccentricity The extent to which the orbit of a planet (such as Earth) departs from a perfect circle.

EMIC (Earth system Model of Intermediate Complexity) A numerical simulation of the evolution of Earth's climate over thousands of years, with simplified geography.

evapotranspiration The process by which trees and other vegetation draw moisture out of the ground and recycle it into the air as water vapor.

fallow The interval between times of cultivation of agricultural plots.

falsification The process through which scientific evidence disproves an idea or hypothesis by contradicting its predictions.

feedback A process internal to Earth's climate system that acts either to amplify changes in climate (positive feedback) or to moderate them (negative feedback).

Fertile Crescent A region in the Middle East encompassing Iraq, Syria, Israel, Jordan, and southern Turkey, in which crops and livestock were domesticated.

forcing Any process or change from previous conditions that drives changes in climate.

forest fallow (also *long fallow* or *slash and burn*) A form of agriculture in which farmers cultivate plots for only a few years, leaving behind abandoned plots where forests slowly regenerate.

forest transition The change observed in many regions (especially in industrialized countries) from long-term deforestation to more recent reforestation.

general circulation model A computer model of the global atmosphere that simulates temperature, precipitation, winds, and atmospheric pressure in three dimensions.

geometric growth Describes an increase in a model parameter (such as population) through time that occurs as a constant fraction of the total amount present (for population, those then living). Also known as *exponential growth*.

girdling Cutting into the outer layers of tree bark to stop the flow of sap and kill the tree.

greenhouse effect The warming of Earth's surface and lower atmosphere that occurs when its own emitted infrared heat is trapped and reradiated downward by greenhouse gases.

greenhouse gases Gases such as carbon dioxide (CO_2) and methane (CH_4) that trap outgoing infrared radiation emitted by Earth's surface, warming the atmosphere.

grid boxes Geometric units within climate models that have uniform climatic characteristics and exchange heat, energy, and other properties with adjoining grid boxes.

hardwood forest Forests of leaf-bearing (deciduous) trees such as oak, maple, ash, hickory, elm, and beech.

hearth counts Population estimates based on the number of households, or fire hearths.

hillforts Large walled enclosures built on the tops of hills thousands of years ago in Europe.

historical archives Sources of information on climate based on human observations of natural phenomena, largely made before the era of instrument measurements.

hydrofracking The injection of pressurized fluids into subsurface rock layers in order to extract natural gas.

hypothesis An explanation of observations based on available principles and evidence.

ice cap A small ice deposit situated on relatively flat topography, usually in the Arctic.

ice flow model A method that simulates ice sheet processes of snow accumulation at higher elevations, internal ice flow, and melting at lower elevations.

ice-rafted debris Sediments of widely ranging sizes eroded from the land by ice, carried to the ocean in icebergs, and deposited on the seafloor.

ice sheets Kilometer-thick mounds of ice situated on continents and their shallow margins.

Industrial Era The interval spanning the last two centuries when power began to be generated by machines driven by the burning of fossil fuels.

inorganic carbon Carbon in mineral form that is not part of the cycle of life.

insolation The amount of solar radiation arriving at the top of Earth's atmosphere, varying by latitude and by season.

interglaciations Intervals of time, lasting less than 10,000 years, when ice sheets were not present on North America or northern Europe.

Iron Age An interval that began near 2,500 years ago when many cultures began to make tools made of iron.

Kutzbach monsoon hypothesis The idea (now a theory) that long-term summer monsoon circulations in the tropics are driven by the amount of incoming solar radiation at tropical/subtropical latitudes.

lichen Primitive vegetation that can grow on trees or rocks as long as sunlight is available.

LIDAR (Light Detection and Ranging) A remote-sensing technology that can penetrate tree canopies and reveal underlying Earth structures, either natural or built.

Little Ice Age An interval between approximately 1400 and 1850 when temperatures in the Northern Hemisphere were generally colder than today, particularly in Europe.

loess Silt-sized, windblown glacial sediment.

logistical growth Describes an increase in some model parameter (for example, population) through time that is initially rapid but then slows as resources become more scarce.

Medieval Warm Period An interval between the years 1000 and 1300 when climate was relatively warm in some parts of Europe.

methane (CH_4) A greenhouse gas that forms in environments that lack oxygen, including natural wetlands and agricultural systems (rice paddies, livestock digestive systems, and burned grasses and crops).

Milankovitch hypothesis The idea that orbitally caused fluctuations in high-latitude solar radiation (insolation) during summer control the size of ice sheets through their effect on melting.

milpa Combined plantings of corn, beans, avocados, and tomatoes in Mexico.

moldboard plow A plow drawn by a team of draft animals, capable of turning heavy sod.

monsoons Seasonally reversing circulations in which winds blow on-shore in summer and offshore in winter because of different rates of heating and cooling of land and water.

mou A standard measure of acreage used historically in China.

mountain glacier A body of ice tens to hundreds of meters thick, and several square kilometers in area, confined to a valley at a high elevation.

Neolithic Agricultural Revolution An interval between 13,000 and 10,000 years ago when agriculture first appeared in the Fertile Crescent, northern China, and parts of South America.

organic carbon Carbon that is part of the cycle of living organisms, occurring most abundantly as vegetation on land and plankton in the ocean.

overkill hypothesis The idea that the sudden extinction of many mammals 12,500 years ago resulted from human hunting rather than climatic stress.

oxidation The addition of oxygen to an element (for example, C + $O_2 \rightarrow CO_2$).

oxygen isotopes Forms of the element oxygen with different atomic masses. Used to trace movement of the heavier (^{18}O) and lighter (^{16}O) forms between the oceans and ice sheets, and to measure changes in ocean temperature.

paleoclimatology The study of past changes in climate and their causes.

pandemic An outbreak of disease at a multicontinental scale.

paradigm shift A change from one generally accepted explanation for observed phenomena to another explanation that better accounts for the evidence.

pastoralism A type of agriculture based on tending livestock that are moved from area to area.

peat bog A deposit of decayed carbon-rich plant remains in a wetland environment.

photosynthesis The process by which plants use nutrients and solar energy to convert water and CO_2 to plant tissue (carbohydrates), thereby producing oxygen.

phytoliths Small fragments of grasses and other plants composed of the compound silica.

phytoplankton Small floating organisms (usually algae) that use energy from the Sun and nutrients from the water for photosynthesis.

plankton Small organisms that float in the upper layers of oceans or lakes.

ppb Parts per billion, the unit commonly used to measure CH_4 concentrations.

ppm Parts per million, the unit commonly used to measure CO_2 concentrations.

precession Changes in Earth's elliptical orbit at cycles of 23,000 and 19,000 years, which causes movement of the seasons through a given year.

radiocarbon dating Dating of relatively young carbon-bearing geologic and archeologic materials by means of ^{14}C, a radioactive isotope that decays by half every 5,700 years.

rain forest Forests composed of leaf-bearing trees that remain evergreen throughout the year because of abundant precipitation and the absence of freezing.

reaves Low stone walls used to mark property boundaries in Britain thousands of years ago.

reforestation The growth of forests in regions that were previously deforested, either by natural processes or by humans.

resolution The degree of detail detected in a climate signal by sampling at a particular interval.

response Changes in the climate system caused by changes in the factors that drive climate.

revolution The once-a-year movement of Earth around the Sun.

rotation The once-a-day spin of Earth around its axis.

savanna A semi-arid region of grasses and scattered trees.

shattering A natural behavior of some annual plants that propels seeds across the surrounding soil area in response to a disturbance.

sickles Crescent-shaped tools designed for cutting grain crops.

soil litter Organic debris (such as leaves, twigs, and dead grass) found at the surface or buried in upper soil layers.

solar radiation Energy emitted by the Sun in visible and ultraviolet wavelengths that delivers heat to Earth's climate system.

stalactites, stalagmites Deposits of calcite ($CaCO_3$) formed in caves by groundwater dripping from above.

steppes Grassy regions with low-lying shrub vegetation, typical of semi-arid areas. In North America, called *prairie*.

stratosphere The stable layer of the atmosphere, lying between 10 and 50 km above Earth's surface, which contains most of Earth's ozone.

sulfate aerosols   Fine particles produced in the atmosphere from sulfur dioxide gas emitted by volcanoes or by industrial smokestacks. These particles can block incoming solar radiation.

sunspots   Dark regions on the Sun's surface caused by the star's internal circulation.

teosinte   A natural plant ancestral to domesticated maize (corn).

terra preta   Soil that has been enriched by addition of charcoal derived from burning vegetation.

theory   A hypothesis that has survived repeated testing and become widely accepted.

tilt   The angle between Earth's equatorial plane and the plane of its orbit around the Sun, also equivalent to the angle between Earth's axis of rotation and a line perpendicular to its axis of rotation around the Sun. Also referred to as *obliquity*.

tree rings   Annual bands formed by trees in regions of seasonal climate, with lighter layers formed during rapid growth in the spring, and darker layers formed at the end of growth in the autumn.

troposphere   The layer of the atmosphere just above Earth's surface (10 km or more thick) in which weather occurs.

tundra   A high-latitude or high-altitude ecosystem in which the ground freezes deeply in winter but thaws at the surface in summer, permitting low-growing plants to flourish.

upwelling   The rise of cool, nutrient-rich subsurface water to the ocean surface to replace warm nutrient-poor surface water.

vegetation-albedo feedback   The process by which changes in vegetation alter the amount of solar radiation absorbed at Earth's surface, amplifying initial temperature changes.

virgin soil pandemic   The large array of diseases brought by Europeans that caused a demographic collapse among Native American populations lacking natural immunity.

Younger Dryas   An interval during the middle of the last deglaciation (near 12,000 years ago) marked by slower melting of ice sheets and a major cooling in the North Atlantic region.

Zebu cattle   Humped cattle domesticated in what is now eastern Pakistan.

INDEX

Note: Page numbers followed by f indicate figures; those followed by t indicate tables.

Adena culture, 111, 112f
aerosols
  biomass burning and, 272–273
  brown clouds and, 272
  cook stoves and, 272
  Industrial Era and, 240, 270–271, 271f
  overview about, 270
  pre-industrial plowing and, 271–272
  solar radiation and, 270
agriculture. See also crop domestication
  anthropogenic deforestation and,
    233–237, 234f, 235f
  axes and, 59, 59f
  Beni province, 109–110, 110f
  Boserup sequence and, 141–145, 141t,
    142f, 143f
  Britain prehistorical example of, 62–65
  Bronze Age and, 46–47, 47f
  child-bearing and, 48
  China and, 45f, 46, 76–79
  deforestation for, 61–62, 71–72, 71f,
    233–237, 234f, 235f
  early crops, 46
  effect on climate, 350, 350f
  Fertile Crescent influence on African,
    119–120, 120f
  Fertile Crescent influence on European,
    58–62, 60f
  Fertile Crescent region of, 45–47, 45f,
    47f, 51, 52t, 54–62, 55f, 56f, 57f,
    59f, 60f, 79, 81f, 119–120, 120f
  historical trends in Europe, 145–148,
    146f, 147f, 148f
  hunter-gatherers and, 47–48
  independent origins and, 48
  on Indian subcontinent, 75t, 79–80
  Iron Age and, 46–47, 47f
  land-use archeology and, 44–45
  long-fallow farming, 61, 141–142,
    141t, 142f, 143f
  Mesoamerica and, 45f, 46,
    99–100, 99f

Neolithic Agricultural Revolution in,
    46–47, 47f, 54–58
  North American centers of early,
    111–113, 111f
  oldest records regarding, 44
  origins of, 44–49, 45f
  plows used in, 63, 63f, 69–70, 69f,
    271–272
  population expansion and, 46–47, 47f,
    48
  radiocarbon dating and, 58, 59f
  slash and burn, 61
  soil fertility and, 60–61
  spread across continents, 134f, 347f
  Stone Age and, 46–47, 47f
  technology improvements, 69–70, 69f
albedo-temperature feedback, 254–255,
    255f, 256f. See also vegetation-
    albedo feedback
  deforestation and, 339–340, 340f
algae, CO_2 levels and, 35–37, 36f
alignment issue. See also
    deglacial-alignment method;
    insolation-alignment method
  orbital changes and, 160–164
  overview about, 159–160
alpaca, 96t
Amazon Basin
  CH_4 increase and, 177–182
  monsoons, 22, 22f, 177–179, 178f
Amazonia
  Beni province in, 109–110, 110f
  crops of, 108, 108f
  deforestation in, 109
  manioc in, 108, 108f
  mortality in, 325, 325f
  terra preta and, 108–109, 109f
  weirs in, 110
Andes
  crops domesticated in, 105–106, 106f
  Inca civilization, 106–108, 323–324,
    324f

Andes (*continued*)
　Incan Peru and, 323–324, 324f
　Norte Chico and, 105
　potatoes and, 106, 106f
　span of, 104–105, 105f
　terracing in, 106, 107f
　Tiwanaku culture, 106
　Wari culture, 106, 107f
animal extinction, 98–99
Antarctica
　EPICA and, 159
　gradual cooling of, 275–276,
　　276f
　greenhouse gases impact on, 279
anthromes, 236–237
Anthropocene. *See also* early
　　anthropogenic hypothesis; Industrial
　　Era Anthropocene paradigm
　climate effects and, 240, 241f, 242
　conclusions about, 242–243
　Crutzen and, 231
　deforestation and, 233–237, 234f,
　　235f
　greenhouse gases and, 237, 238f,
　　239–240
　Industrial Era transformations and,
　　231–232, 232t
　Stoermer and, 231
　temperature and, 240, 241f, 242
　theory established, 231
　time regarding, 240
Antonine Plague, 313–314
Arctic
　CH_4 and, 24–25, 24f, 26f
　gradual cooling of, 276–277, 276f,
　　277f
　wetlands, 24–25, 24f
Australia
　conclusions about, 132
　early climate change in, 129
　fire and, 204
　firesticks used in, 128–129
　marsupials in, 128–129
　overview about, 117
　vegetation groups in, 128, 128f
axes, 59, 59f
axis tilt, 6, 7f
Aztec Mexico, 322–323, 323f

Baffin Island, 292, 293f, 294
Bantu people, 125–126, 126f, 127–128,
　127f
beans, 100, 100f
　mung, 79, 82f
Beni province, 109–110, 110f
Bern group, 195, 196f, 197
biomass burning, 272–273
biomes, 236–237
Black Death, 315–317, 316f
Boserup, Ester, 141
Boserup sequence
　conclusions about, 144–145
　double cropping and, 144
　farming innovation and, 143–144
　labor and, 144
　long-fallow farming and, 141–142,
　　141t, 142f, 143f
　overview about, 141, 141t
　smaller plots shift and, 143
Britain
　deforestation in, 65–66, 66f, 68,
　　71–72, 71f
　Domesday Book survey of England
　　and Wales and, 145, 146f
　hillforts in, 63, 64f, 65
　plague in, 68
　plows used in, 63, 63f
　prehistorical example of agriculture,
　　62–65
　round houses in, 65
　sediment influx in, 63, 64f
Broecker, Wallace, 190, 200
Bronze Age
　agriculture and, 46–47, 47f
　in China, 87
brown clouds, 272
bubonic plague
　Black Death, 315–317, 316f
　CO_2 concentration changes and, 313f
　described, 315
　last significant outbreak of, 317
　overview, 312, 313f
　Plague of Justinian, 314–315
　second wave of, 315–317,
　　316f
Buck, John, 90, 148–149, 149f
buffalo, 114

Cahokia mounds, 113, 113f
Caracol (Mayan city), 103, 103f
carbon dioxide (CO_2)
  algae and, 35–37, 36f
  anthropogenic emissions feedbacks, 197–199, 198f, 199f
  anthropogenic role in short-term decreases in, 331–338, 332f, 335f, 341–342, 341f
  archeological data and, 207, 349
  Asian early trends of, 92–93
  assessment of alignment methods and, 168–171, 169f
  Bantu people and, 127–128
  bubonic plague and, 313f
  carbon emissions mismatch with concentrations of, 201–205, 201f, 202f
  carbon reservoirs and, 31–33, 32f
  carbonate compensation hypothesis and, 190–192, 191f
  carbon/climate models and, 306, 307f
  carbon-isotope evidence, 195, 196f, 197
  CH_4 compared with, 31
  changes in anthropogenic sources of, 193t
  changes in natural sources of, 190t
  China and, 92–93, 318, 318f
  comparisons regarding, 208
  conclusions about, 205, 206–208, 207t
  coral reef hypothesis and, 192–193, 192f
  deep ocean and, 35–37, 35f, 36f
  deglacial-alignment method and, 166–167, 166f, 167f, 168t
  early Americas and, 115
  early anthropogenic hypothesis and, 250–251, 252f, 253
  falsification and, 217–219, 217t, 218f
  fertilization, 306
  fire and, 204–205
  40-ppm proposed anthropogenic sources of, 193–195, 194f, 196f, 197–199, 198f, 199f
  fossil fuels and, 280–281
  frost-free season and, 306
  future changes in, 279–280, 280f, 281f

  geometric growth assumption and, 202–203, 202f
  glaciations and missing, 33–37
  historical data and, 207
  The Holocene and, 229
  ice-core chemistry and, 11–13, 13f, 32–33, 33f
  ice volume variations and, 32–33, 33f, 34f
  ice-age cycles and, 2–4, 3f
  Industrial Era Anthropocene paradigm and, 237, 238f, 239
  insolation-alignment method and, 162, 163f, 164t
  Law Dome records of, 303, 304f
  logistical growth assumption and, 202f, 203
  models and, 207–208
  mortality overview and, 311, 312f, 351, 352f
  natural explanations for decreases in, 303–307, 305f, 307f
  no future glaciations and, 279–283, 280f, 281f, 282f
  ocean temperature and, 304, 305f
  oxygen-isotope index and, 198
  population and, 202–204, 202f
  present-day storage of, 336, 337f
  previous interglaciations and, 39–40, 162, 163f, 164t
  reforestation and, 334
  short-term decreases of, anthropogenic role in, 331–338, 332f, 335f, 341–342, 341f
  soil litter and, 304–305, 305f
  sources tests, 200–201
  Southern Ocean and, 199, 199f
  stable interglacial climate and, 38–39
  temperature and, 197–198, 198f
  trends over last 2000 years, 288, 289f
  22-ppm increase proposed sources, 189–193, 190t, 191f, 192f
  unexpected level changes in, 37–39, 38f
  virgin soil pandemic and, 321f
  wrong-way trend in, 31–42, 42f, 345, 346f

carbon emissions
  CO_2 concentration mismatch with, 201–205, 201f, 202f
  conclusions about, 205, 206–208, 207t
  fire and, 204–205
  geometric growth assumption and, 202–203, 202f
  logistical growth assumption and, 202f, 203
  population trends and, 202–204, 202f
carbon reservoirs, 31–33, 32f
  photosynthesis and oxidation and, 32
carbonate compensation hypothesis, 190–192, 191f, 200
carbon/climate models, 306, 307f
carbon-isotope evidence, 195, 196f, 197
Carcaillet, Christopher, 193
Carthaginian Empire, 119
CCSM. *See* community climate science model
Central America
  milpa in, 101–102
  squash domesticated in, 100, 100f
  teosinte and, 101, 101f
CH_4. *See* methane
Chappelaz, Jérôme, 183
charcoal, 108–109, 109f, 130
chickpea domestication, 55, 55f
China
  agriculture and, 45f, 46, 76–79
  archeological sites in, 77, 78f
  Bronze Age in, 87
  civil strife in, 317–320, 318f, 319f
  CO_2 and CH_4 early trends in, 92–93
  CO_2 concentration changes and, 318, 318f
  CO_4 emissions in early, 89–90, 92–93
  coal use in, 92, 197
  crop domestication in, 75t
  crop fertilization in, 88–89
  deforestation in, 87, 92
  farming tools in, 76, 77f
  fruit tree grafting in, 87
  historical record in, 88–92
  land division practices in, 91
  land use trends in, per capita, 148–149, 149f, 150f, 151
  land-use change in, 90, 90f

  livestock domestication in, 75t, 76–79
  livestock spread into, 80–82, 83f, 84
  Manchu conquest in, 320
  millet farming in, 76, 76f
  Mongol warriors and, 318–319, 319f
  mortality and, 317–320, 318f, 319f
  phytoliths and, 86
  pigs and, 76, 76f
  pollen records in, 85–86
  population increasing in, 77–79
  rice as dryland crop in, 84, 84f
  rice as irrigated crop in, 84–88
  rice farming in, 76
  rice paddy fields in, 85, 85f, 87, 89
  rice terraces in, 89, 89f
  smallpox in, 320
  spread of rice in, 86–87, 86f
  Yangtze River basin rice irrigation in, 91, 91f
climate engineering, 283–284
climate proxies
  ice cores, 295–296, 296f, 297
  Little Ice Age and, 295–298, 296f, 297f
  temperature estimates from, 295–298, 296f, 297f
  tree rings, 295, 296–297, 296f
  CO_2. *See* carbon dioxide
coal
  charcoal, 108–109, 109f, 130
  China and, 92, 197
  falsification and, 218
community climate science model (CCSM)
  albedo-temperature feedback and, 254–255, 255f, 256f
  general circulation model and, 254
  resolution, 259–260, 260f
  summer snow cover and, 256–257, 258f, 259, 350–351, 351f
  winter sea ice extent and, 256, 257f
The Conditions of Agricultural Growth (Boserup), 141
conifers, 264, 265f
constant land use assumption, 140–141
  Buck and, 148–149, 149f

decreasing per capita land use significance, 154–155, 155f
  Kaplan and, 145–147, 147f, 148f, 149, 150f, 151
  Mather and, 147
  Rackham and, 145
  rice irrigation and, 151–154, 152f, 154f
cook stoves, 272
cooling trend, 275–277, 276f, 277f
coral reef hypothesis, 192–193, 192f, 200
Cortéz, Hernán, 322, 323f
Cro-Magnon people, 52–53
crop domestication
  African, 119t, 121, 125, 125f, 126f
  Amazonia, 108, 108f
  American, 96t
  Andean, 105–106, 106f
  Asian, 75t, 76–79, 133f
  chickpea, 55, 55f
  early, 46
  Fertile Crescent, 51, 52t, 54–58, 55f, 56f, 57f
  flax, 55–56
  in India, 75t, 79–80
  Mesoamerican, 133f
  rice, 84–88
  summary, 133–134
  timeline, 133f
  in tropical Africa, 125, 126f
  wheat, 55, 55f
Crosby, Alfred, 320
Crutzen, Paul, 231

de Soto, Hernando, 113, 325–326, 326f
deforestation
  for agriculture, 61–62, 71–72, 71f, 233–237, 234f, 235f
  albedo changes due to, 339–340, 340f
  in Amazonia, 109
  anthromes and, 236–237
  anthropogenic, 233–237, 234f, 235f
  atmospheric dust and, 270–273, 271f
  Boserup sequence and, 141–145, 141t, 142f, 143f
  in Britain, 65–66, 66f, 68, 71–72, 71f
  Buck and, 148–149, 149f
  in China, 87, 92
  early anthropogenic hypothesis and, 233, 234f, 348–349, 348f
  in Europe, 61–62, 65–67, 66f, 71–72, 71f, 145–148, 146f, 147f, 148f
  evapotranspiration and, 266, 266f
  false assumption of, 235–237
  fire and, 204–205
  He and, 269, 270f
  high northern latitudes and, 264–265, 265f
  historical evidence of, 235–236
  historical trends in Europe, 145–148, 146f, 147f, 148f
  incoming solar radiation effects of, 263–266, 264f, 265f, 266f
  Industrial Era view of, 233, 234f, 348–349, 348f
  Kaplan and, 145–147, 147f, 148f, 149, 150f, 151
  land-use records and, 233, 235f
  Mather and, 147
  middle northern latitudes and, 265
  model simulations about, 269–270, 270f, 349
  modeling experiments on, 234f, 236
  as negative, 71–72
  New Zealand, 131
  Pongratz and, 269
  population trends and, 202–204, 202f
  as positive, 71
  predicted effects of pre-industrial, 267–268, 267f
  Rackham and, 145
  reforestation and, 233–235, 235f
  timing of European, 71f
  two thousand years ago, 267f
  two views of, 233, 234f
  vegetation-albedo feedback and, 264–265, 265f
deglacial-alignment method
  assessment of, 168–171, 169f
  conclusions of, 167
  greenhouse gases and, 164–168, 165f, 166f, 167f, 168t
  interglacial stage 11 and, 171–173, 172f
  interglaciation gas trends using, 166–167, 166f, 167f, 168t

deglacial-alignment method (*continued*)
  overview about, 164–165
  oxygen-isotope index and, 165–166,
    165f
Denevan, William, 320
disciplinary matrix, 221
Dobyns, Henry F., 322
Domesday Book survey of England and
  Wales, 145, 146f
domestication. *See* crop domestication;
  livestock domestication
double cropping, 144

early anthropogenic hypothesis
  agriculture effect on climate and, 350,
    350f
  alignment issue regarding, 159–160
  anthromes and, 236–237
  assessment of alignment methods and,
    168–171, 169f
  Boserup sequence and, 141–145, 141t,
    142f, 143f
  Buck and, 148–149, 149f
  CH_4 and, 250–251, 253
  climate effects and, 240, 241f, 242
  CO_2 and, 250–251, 252f, 253
  conclusions about, 205, 284–285
  decreasing per capita land use signifi-
    cance and, 154–155, 155f
  deforestation view of, 233, 234f,
    348–349, 348f
  Domesday Book and, 145, 146f
  EMIC and, 253
  general circulation model and,
    253–257, 257f, 258f, 259–260
  glacial inception and, 250–251, 252f,
    253
  greenhouse-gas emissions and, 237,
    238f, 239–240, 250–251, 252f, 253
  historical trends in Europe and,
    145–148, 146f, 147f, 148f
  The Holocene and, 229
  insolation trend and, 250, 251f
  interglacial stage 11 and, 171–173,
    172f
  Kaplan and, 145–147, 147f, 148f, 149,
    150f, 151
  Mashey and, 226

Mather and, 147
model simulation types, 253–254
Nature and, 226
overview about, 136–137
paradigm shift and, 212
population and, 139, 140f
process leading to, 224–225
Rackham and, 145
resistance to, 227–228
rice irrigation and, 151–154, 152f,
  154f
Science and, 226–227
small population and, 139, 140f
temperature and, 240, 241f, 242
earthen mounds
  Cahokia mounds, 113, 113f
  North American, 112f, 113,
    113f
  Serpent Mound, 113, 114f
Earth's Climate: Past and Future
  (Ruddiman), 225
Earth's orbit
  alignment issue and, 160–164
  axis tilt and, 6, 7f
  changes in, 6–8, 7f, 8f
  insolation and, 7–8, 8f
  path changes, 6, 7f
  summer monsoons and, 13–16
  wobble of, 7, 7f
Earth system model of intermediate
  complexity (EMIC), 253
East African Highlands, 123,
  124f
eccentricity, 6, 7, 7f, 171
Egypt, 119–120, 121f
Ellis, Erle, 236–237
EMIC. *See* Earth system model of
  intermediate complexity
emissions feedbacks, 197–199, 198f,
  199f
EPICA. *See* European Project for Ice
  Coring in Antarctica
Eric the Red, 292
European Project for Ice Coring in
  Antarctica (EPICA), 159
evapotranspiration, 266, 266f
evergreen conifers, 264, 265f
extinction, animal, 98–99

falsification, 211
  CH$_4$ trend and, 214–217, 215f, 216t
  CO$_2$ trend and, 217–219, 217t, 218f
  coal and peat and, 218
  conclusions about, 219
  as not absolute, 213–214
  Popper and, 213
famine, 317
farming. *See* agriculture
feedback. *See* albedo-temperature
    feedback; emissions feedbacks;
    vegetation-albedo feedback
Fertile Crescent, 45, 45f, 54f
  Africa and, 119–120, 120f
  axes and, 59, 59f
  chickpea domestication in, 55, 55f
  crop and livestock domestication in,
    51, 52t, 54–58, 55f, 56f, 57f
  European agriculture and, 58–62, 60f
  flax domestication in, 55–56
  India and, 79, 81f
  Neolithic Agricultural Revolution in,
    46–47, 47f, 54–58
  pathways into Europe, 60, 60f
  radiocarbon dating regarding, 58, 59f
  seed scattering and, 54
  seed shattering and, 54–55
  tools, 56, 56f
  wheat domestication in, 55, 55f
fertilization, CO$_2$, 306
fire
  Australia and, 128–129, 204
  carbon emissions and CO$_2$ mismatch
    and, 204–205
  Maori people and, 204
The Fire Eaters (Flannery), 128–129
Flannery, Tim, 128–129
flax domestication, 55–56
forest clearance. *See* deforestation
forest fallow farming. *See* long-fallow
    farming
forest transition, 233
forest vegetation
  evergreen conifer, 264, 265f
  natural global, 264f
  solar radiation and, 263–264, 264f
  tundra and, 264, 265f
fossil fuels, 280–281

Four Horsemen of the Apocalypse, 315,
    316f
fracking, 280
frost-free season, 306
fruit tree grafting, 87
Fuller, Dorian, 183, 239

gazelles, 56
general circulation model
  albedo-temperature feedback and,
    254–255, 255f, 256f
  CCSM and, 254
  conclusions, 260
  early anthropogenic hypothesis and,
    253–257, 257f, 258f, 259–260
  overview about, 253–254
  positive feedback effects and,
    254–255, 255f, 256f
  resolution, 259–260, 260f
  summer snow cover and, 256–257,
    258f, 259, 350–351, 351f
  using, 254–260
  winter sea ice extent and, 256, 257f
Genghis Khan, 319, 319f
geometric growth assumption, 202–203,
    202f
girdling, 59, 59f
glaciation
  missing CO$_2$ and, 33–37
  no future, 279–283, 280f, 281f, 282f
glaciation, next
  EMIC and, 253
  general circulation model and,
    253–257, 257f, 258f, 259–260
  greenhouse-gas emissions and,
    250–251, 252f, 253
  insolation trend and, 250, 251f
  Milankovitch hypothesis and,
    249
  model simulation types and, 253–254
  natural conditions and, 277–278
  as not happening, 279–283, 280f,
    281f, 282f
  as overdue, 250, 251f
  skipped beat, 278–279, 278f
  threshold for, 282, 282f
glacier formation, 259
Glacier National Park, 295f

grafting, fruit tree, 87
greenhouse gases. *See* carbon dioxide; methane
Greenland
  ice sheet, 279
  Little Ice Age and, 292
  Norse settlement on, 292
Gregg, Susan, 61–62
Gupta era, 88

Harappan civilization, 79, 80f
Hays, Jim, 249
He, Feng, 269, 270f
high northern latitudes, 264–265, 265f
hillforts, 63, 64f, 65
The Holocene (journal), 229
Hopewell culture, 111–112, 112f
hunter-gatherers, 47–48, 52–53, 345
hunting and gathering
  DNA studies and, 52
  early prehistorical record of, 52–53
hydrofracking, 280

ice-age cycles. *See also* glaciation, next; interglaciations
  Antarctica, 275–276, 276f
  Arctic, 276–277, 276f, 277f
  skipped beat, 278–279, 278f
  trends in, 2–4, 3f
  vegetation reductions during, 34
ice caps, 292, 293f
ice-core
  as climate proxy, 295–296, 296f, 297
  EPICA and, 159
ice-core chemistry
  CH_4 and, 11–13, 13f
  CO_2 and, 11–13, 13f, 32–33, 33f
ice sheet changes. *See also* glaciation, next; interglaciations
  Antarctica, 275–276, 276f
  Arctic, 276–277, 276f, 277f
  CO_2 and, 32–33, 33f, 34f
  conclusions about, 16–18
  ice-core chemistry and, 11–13, 13f
  insolation and, 8–13
  Milankovitch hypothesis and, 11
  North American, 39f

ocean sediments and, 9–10, 10f
  over millions of years, 275–277, 276f, 277f
  oxygen isotopes and, 9–10
  skipped beat, 278–279, 278f
  in summer northern hemisphere, 11, 12f
Iceland, Little Ice Age and, 291, 292f
Imbrie, John, 249
Inca civilization, 106–108, 323–324, 324f
Incan Peru, 323–324, 324f
incremental research, 221
incremental science, 210
India
  crop domestication in, 75t, 79–80
  Fertile Crescent crops spread into, 79, 81f
  Gupta era in, 88
  Harappan civilization in, 79, 80f
  historical record in, 88
  livestock domestication in, 75t, 79–80
  livestock spread into, 80–82, 83f, 84
  Mauryan civilization in, 88
  mung beans and, 79, 82f
  rice spread into, 87
  Zebu cattle in, 79, 81, 82f, 83f
Industrial Era
  atmospheric dust and, 240, 270–271, 271f
  deforestation and, 233, 234f, 348–349, 348f
  environmental consequences of, 232t
  innovations of, 231–232, 232t
  solar radiation and, 270–271, 271f
Industrial Era Anthropocene paradigm
  anthromes and, 236–237
  atmospheric dust and, 240
  CH_4 and, 237, 239–240
  climate effects and, 240, 241f, 242
  CO_2 and, 237, 238f, 239
  conclusions about, 242–243
  criticisms of, 233–236, 235f
  deforestation false assumption of, 235–237
  deforestation view of, 233, 234f, 348–349, 348f

greenhouse-gas emissions and, 237,
  238f, 239–240
land-use records and, 233, 235f
reforestation and, 233–235, 235f
temperature and, 240, 241f, 242
time regarding, 240
innovative science, 210
insolation
  changes in northern hemisphere
    summer, 11, 12f
  conclusions about, 16–18
  ice-core chemistry and, 11–13, 13f
  ice sheet changes and, 8–13
  interglaciations and, 17–18, 17f, 18t
  Kutzbach monsoon hypothesis and,
    15–16, 15f, 16f
  lake-level evidence and, 22–23, 23f
  last fifteen thousand years and, 19, 21f
  Little Ice Age and, 298, 298f
  Milankovitch hypothesis and, 11
  Northern and Southern Hemispheres
    and, 177–178, 178f
  orbital cycles and, 7–8, 8f
  over last 130,000 years, 180–181,
    181f
  overdue glaciation and, 250, 251f
  trend in summer, 250, 251f
  wetlands and, 22–24, 23f
insolation-alignment method
  ages of interglaciations and, 161t
  assessment of, 168–171, 169f
  CH_4 concentrations and, 160, 162,
    162f, 163t
  CO_2 concentrations and, 162, 163f,
    164t
  greenhouse gases and, 160–164, 161f,
    161t, 162f, 163f, 163t, 164t
  interglacial stage 11 and, 171–173
  solar radiation changes and, 161f
  steps in, 160
interglaciations
  ages of, 161t
  anthropogenic explanation and, 200
  carbonate compensation hypothesis
    and, 200
  CH_4 and, 27–28, 27f, 160, 162, 162f,
    163t
  CO_2 and, 38–40, 162, 163f, 164t

coral reef hypothesis and, 200
deglacial-alignment method and,
  166–167, 166f, 167f, 168t,
  171–173, 172f
EPICA and, 159
insolation and, 17–18, 17f, 18t
longevity of, 278–279
stage 11, 171–173, 172f
summer solar radiation changes and,
  161f
Inuits, 292
Iron Age agriculture, 46–47, 47f
isotopes, oxygen, 9–10

Kaplan, Jed, 200, 236
  deforestation and, 145–147, 147f,
    148f, 149, 150f, 151
  40-ppm proposed anthropogenic
    sources and, 194, 194f
  short-term CO_2 decreases and,
    334–336, 335f
Kaufman, Darryl, 298, 298f
Keeling, Charles, 249–250
Kerr, Dick, 227
Khans, 319, 319f
Khoisan people, 127
Kuhn, Thomas
  largest paradigm shifts and, 223–224
  normal science and, 221–222
  novel ideas and, 223
  The Structure of Scientific Revolutions
    and, 221
Kutzbach, John, 15–16, 15f, 16f, 254
Kutzbach monsoon hypothesis, 15–16,
  15f, 16f

lake-level evidence, 22–23, 23f
land use. *See also* constant land use
    assumption
  archeology, 44–45
land use, per capita
  Boserup sequence and, 141–145, 141t,
    142f, 143f
  conclusions about, 156
  constant land use assumption and,
    140–141
  by early farmers, 142–143
  early farming and, 139–156

land use, per capita (*continued*)
  historical trends in China, 148–149, 149f, 150f, 151
  historical trends in Europe, 145–148, 146f, 147f, 148f
  rice irrigation and, 151–154, 152f, 154f
  significance of decreasing, 154–155, 155f
Law Dome, 303, 304f
least publishable units (LPUs), 222
lichen, 292, 293f, 294
LIDAR. *See* Light detection and ranging
Light detection and ranging (LIDAR), 103, 103f
Little Ice Age, 288
  anthropogenic role regarding, 341–342, 341f
  Baffin Island and, 292, 293f, 294
  carbon/climate models and, 306, 307f
  CH_4 and, 308–309
  climate proxies and, 295–298, 296f, 297f
  conclusions about, 301–302
  Europe and, 295
  frost-free season and, 306
  Greenland and, 292
  ice caps and, 292, 293f
  Iceland sea ice and, 291, 292f
  insolation and, 298, 298f
  Kaufman and, 298, 298f
  Law Dome records of, 303, 304f
  lichen and, 292, 293f, 294
  mountain glaciers and, 292, 293f, 294f, 295
  natural explanations for, 298–301, 298f, 299f, 300f, 301f, 303–307, 305f, 307f
  observations of, 291–295
  ocean temperature and, 304, 305f
  overview about, 291
  sea ice and, 291–292, 292f
  soil litter and, 304–305, 305f
  solar radiation and, 298–301, 298f, 299f, 300f, 301f
  temperature and, 295, 304, 305f
  volcanic eruptions and, 299, 299f, 300f
livestock domestication
  in Africa, 119t, 123f

  alpaca, 96t
  in Americas, 96t
  in Asia, 75t, 76–79
  CH_4 increase and, 183–185, 186f
  in China, 75t, 76–79
  Fertile Crescent, 51, 52t, 54–58, 55f, 56f, 57f
  on Indian subcontinent, 75t, 79–80
  llama, 96t
  summary, 133–134
  Yangtze River basin and, 91–92
llama, 96t
logistical growth assumption, 202f, 203
long-fallow farming, 61, 141–142, 141t, 142f, 143f

maize
  domesticated in Mesoamerica, 100
  teosinte and, 101, 101f
Manchu conquest, 320
manioc, 108, 108f
Maori people, 129–130
  fire and, 204
marsupials, 128–129
Mashey, John, 226
Mather, Alexander, 147
Mauryan civilization, 88
Mayan civilization, 103–104, 103f
McNeill, William, 312
Medieval Warm Period, 288
Mediterranean, 118–119, 120f
  mortality, mass human, and, 312–317, 313f
Mesoamerica. *See also* Central America; Mexico
  agriculture and, 45f, 46, 99–100, 99f
  beans domesticated in, 100, 100f
  crops domestication timeline in, 133f
  maize domesticated in, 100
  Mayan civilization in, 103–104, 103f
  milpa in, 101–102
  Olmec civilization in, 102–103
  spread of crops in, 101–102, 102f
  squash domesticated in, 100, 100f
  Tenochtitlan in, 104, 104f
methane (CH_4)
  anthropogenic role in short-term decreases in, 338–342, 340f, 341f

anthropogenic sources of, 176t,
    182–185, 184f, 185f, 186f
archeological data and, 207, 349
Arctic and, 24–25, 24f, 26f
assessment of alignment methods and,
    168–171, 169f
biomass burning and, 185
changes in anthropogenic sources of,
    176t
changes in natural sources of, 175t
China, trends in early, 92–93
CO$_2$ compared with, 31
comparisons regarding, 208
conclusions about, 186
decreases effect on climate, 341–342
deglacial-alignment method and,
    166–167, 166f, 167f, 168t
drying and, 177
early Americas and, 115
early anthropogenic hypothesis and,
    250–251, 253
emissions, in early China, 89–90
falsification and, 214–217, 215f, 216t
full anomaly of, 183, 185f
Fuller study and, 183, 184f
historical data and, 207
The Holocene and, 229
human waste and, 185
ice-age cycles and, 2–4, 3f
ice-core chemistry and, 11–13, 13f
Industrial Era Anthropocene paradigm
    and, 237, 239–240
insolation and, 22–24, 23f, 160, 162,
    162f, 163t
insolation-alignment method and, 160,
    162, 162f, 163t
lake-level evidence and, 22–23, 23f
last 130,000 years and, 180–181, 181f
last fifteen thousand years and, 19, 21f
Law Dome records of, 303, 304f
Little Ice Age and, 308–309
livestock and, 183–185, 186f
models and, 207–208
from monsoon wetlands, 21–24, 22f, 23f
mortality overview and, 311, 312f,
    351, 352f
natural explanations for decreases in,
    308–309

natural sources of, 176–177, 176f
north–south concentration
    differences, 25–27, 26f
precipitation and, 309
previous interglaciations and, 27–28,
    27f, 160, 162, 162f, 163t
production of, 19, 20f
rice paddies and, 87, 89
river-mouth deltas as source of, 182
short lifetime in atmosphere of, 25
Singarayer study on, 180–182, 181f
South America and, 177–182
temperature and, 309
tropics and, 25–26, 26f
unexpected increase in, 23–24
wetlands and, 20f, 25–27, 26f,
    176–177, 176f
wrong-way trend in, 19–28, 41–42,
    42f, 345, 346f
Younger Dryas and, 19
Mexico
    Aztec, 322–323, 323f
    beans domesticated in, 100, 100f
    maize domesticated in, 100
    Mayan civilization in, 103–104, 103f
    milpa in, 101–102
    Olmec civilization in, 102–103
    Tenochtitlan in, 104, 104f
    teosinte and, 101, 101f
Milankovitch hypothesis, 11, 249
millet farming, 76, 76f
milpa, 101–102
Mississippian culture, 112–113, 112f,
    113f, 326–327, 326f
moldboard plow, 69–70, 69f
Mongol warriors, 318–319, 319f
monsoons
    African, 22, 22f
    Amazon Basin, 22, 22f, 177–179, 178f
    Asian, 21–22, 22f
    assessment of alignment methods and,
        168–171, 169f
    CH$_4$ from wetlands during, 21–24,
        22f, 23f
    conclusions about, 16–18
    Earth's orbit and, 13–16
    factors enhancing, 21
    insolation affecting, 22–24, 23f

monsoons (*continued*)
  Kutzbach hypothesis of, 15–16, 15f, 16f
  lake-level evidence and, 22–23, 23f
  mechanisms of, 13–14, 15f
Monte Verde, 98
mortality, mass human
  Amazonia, 325, 325f
  Aztec Mexico, 322–323, 323f
  bubonic plague and, 312–317, 313f
  China and, 317–320, 318f, 319f
  Europe and, 312–317, 313f
  Incan Peru, 323–324, 324f
  Mediterranean and, 312–317, 313f
  North America, 325–328, 326f, 327f
  overview, 311, 312f, 351, 352f
  reforestation and, 331–333, 332f
  virgin soil pandemic, 320–328
Mount Pinatubo, 299, 299f
mountain glaciers, 288
  Little Ice Age and, 292, 293f, 294f, 295
  Thunderbird Glacier, 295f
mung beans, 79, 82f

Nature (journal), 226
Neanderthals, 52–53
Neolithic Agricultural Revolution, 46–47, 47f, 54–58
New Guinea, 129–130, 129t
New Zealand
  conclusions about, 132
  fire and, 204
  human impact on, 130–131, 131f
Nile River, 119–120, 121f
nomadic groups, 120
normal science, 221–222
Norse settlement, on Greenland, 292
Norte Chico, 105
North America
  Adena culture in, 111, 112f
  Cahokia mounds in, 113, 113f
  centers of early agriculture in, 111–113, 111f
  de Soto and, 113, 325–326, 326f
  earthen mounds in, 112f, 113, 113f
  eastern, 114, 327, 327f
  Hopewell culture in, 111–112, 112f

  ice sheet changes, 39f
  Mississippian culture in, 112–113, 112f, 113f, 326–327, 326f
  mortality in, 325–328, 326f, 327f
  Serpent Mound in, 113, 114f

ocean
  carbonate compensation hypothesis and, 190–192, 191f
  coral reef hypothesis and, 192–193, 192f
  deep, 35–37, 35f, 36f
  feedbacks and, 197–199, 198f, 199f
  sediments, 9–10, 10f
  Southern Ocean, 199, 199f
  temperature and, 197–198, 198f, 304, 305f
Olmec civilization, 102–103
overkill hypothesis, 98–99
oxidation, 32
oxygen-isotope index, 198
  deglacial-alignment method and, 165–166, 165f
oxygen isotopes, 9–10

paddy fields, rice, 85, 85f, 87, 89
Pakistan, 79, 80f
paleoclimatology
  evolution of, 5–6
  public awareness and, 5
pandemics, 288
paradigm. *See also* Industrial Era Anthropocene paradigm
  acceptance of new, 222–224
  defined, 221
  hypotheses and, 225–226
  normal science and, 221–222
  novel ideas and, 223
  puzzle solving and, 222
paradigm shift
  anomalous results and, 222
  Kuhn and, 221–224
  largest shifts, 223–224
  Plank on, 223
  platetectonic hypothesis as, 223
  science and, 211–212
  time and, 223
peat, 195, 196f, 197, 218
Peru, Incan, 323–324, 324f

photosynthesis, 32
phytoliths, 86
pigs, 76, 76f
Pinatubo, Mount, 299, 299f
Pizarro, Francisco, 323–324, 324f
Pizarro, Hernando, 323–324, 324f
plague, 67–68. *See also* Antonine Plague;
  bubonic plague
  Plows, Plagues, and Petroleum,
    224, 227
Plague of Cyprian, 314
Plague of Justinian, 314–315
Plagues and Peoples (McNeill), 312
Plank, Max, 223
platetectonic hypothesis, 223
plow
  Britain and, 63, 63f
  moldboard, 69–70, 69f
  pre-industrial, 271–272
Plows, Plagues, and Petroleum
  (Ruddiman), 224, 227
pollen records, 85–86
Pongratz, Julia, 269
Popper, Karl, 213
population. *See also* mortality, mass human
  agriculture and, 46–47, 47f, 48
  carbon emissions and CO_2 mismatch
    and, 202–204, 202f
  child-bearing and, 48
  China's early rising, 77–79
  deforestation and, 202–204, 202f
  early anthropogenic hypothesis and,
    139, 140f
  European trends in, 67–69
  geometric growth assumption and,
    202–203, 202f
  logistical growth assumption and,
    202f, 203
  trends in, 47f, 140f, 202–204, 202f
Population and Technological Change
  (Boserup), 141
positive feedback effects, 254–255, 255f,
  256f
potatoes, 106, 106f
precession, 7, 7f, 16, 169
precipitation
  CH_4 and, 309
  tree rings and, 296–297

Rackham, Oliver, 145
radiocarbon dating, 58, 59f
rain forests, 265
  evapotranspiration and, 266, 266f
Red, Eric the, 292
reforestation
  CO_2 storage variables and, 334
  Industrial Era Anthropocene paradigm
    and, 233–235, 235f
  mass mortality and, 331–333, 332f
  quantifying impacts of, 333–338
rice
  archeological evidence regarding,
    153–154, 154f, 349
  CH_4 and, 87, 89
  domestication and irrigation of,
    84–88
  as dryland crop, 84, 84f
  farming, 76
  Fuller study regarding, 183, 184f
  historical evidence regarding,
    151–153, 152f
  historical record regarding, 88
  as irrigated crop, 84–88
  irrigation land use changes, 151–154,
    152f, 154f
  paddy fields, 85, 85f, 87, 89
  phytoliths and, 86
  pollen records related to, 85–86
  shattering and, 85
  spread into India, 87
  spread of, in China, 86–87, 86f
  terraces, 89, 89f
  Yangtze River basin irrigation of, 91,
    91f
Ridgwell, Andrew, 192, 200
river-mouth deltas, 182
Roman Europe, 313–314
round houses, 65

Saharan North Africa, 118–120, 120f
Sahel, 121, 122f, 123, 124f, 125f
savanna, 121, 122f, 123, 124f, 125f
Schneider, Stephen, 213
science
  advancement types, 210
  anomalous results and, 222
  conclusions about, 230

science (*continued*)
  confrontational approach to, 213
  falsification and, 211
  incremental, 210
  innovative, 210
  Kuhn and, 221–222
  normal, 221–222
  paradigm shift and, 211–212
  Plank on, 223
  Popper and, 213
Science (journal), 226–227
Science as a Contact Sport (Schneider), 213
Serpent Mound, 113, 114f
Shackleton, Nick, 249
shattering
  Fertile Crescent region and, 54–55
  rice and, 85
Singarayer, Joi, 180–182, 181f
skipped beat, 278–279, 278f
slash and burn agriculture, 61
smallpox, 320, 322, 324
soil litter, 304–305, 305f
solar radiation
  atmospheric dust and, 270
  biomass burning and, 272–273
  brown clouds and, 272
  cook stoves and, 272
  deforestation and, 263–266, 264f, 265f, 266f
  evapotranspiration and, 266, 266f
  evergreen conifers and, 264, 265f
  forest vegetation and, 263–264, 264f
  high northern latitudes and, 264–265, 265f
  ice-age cycles and, 2–4, 3f
  Industrial Era and, 270–271, 271f
  insolation-alignment method and, 161f
  Little Ice Age and, 298–301, 298f, 299f, 300f, 301f
  middle northern latitudes and, 265
  pre-industrial plowing and, 271–272
  rain forests and, 265–266
  tundra and, 264, 265f
  variations, 299–301, 301f
  vegetation-albedo feedback and, 264–265, 265f
  volcanic eruptions and, 299, 299f, 300f

sorghum, 121, 125f
Southern Ocean, 199, 199f
Stoermer, Eugene, 231
Stone Age agriculture, 46–47, 47f
The Structure of Scientific Revolutions (Kuhn), 221
subtropical southern Africa, 126–128
summer snow cover, 256–257, 258f, 259, 350–351, 351f

taro, 124f, 125, 126f, 127, 130, 130t, 134
temperature
  albedo-temperature feedback and, 254–255, 255f, 256f, 339–340, 340f
  carbon/climate models and, 306, 307f
  CH_4 and, 309
  climate proxies and, 295–298, 296f, 297f
  CO_2 and, 197–198, 198f, 304, 305f
  40-ppm proposed anthropogenic sources and, 197–198, 198f
  future changes in, 279, 281f
  Industrial Era Anthropocene paradigm and, 240, 241f, 242
  Little Ice Age and, 295, 304, 305f
  ocean, 197–198, 198f, 304, 305f
  time regarding, 240
  tree rings and, 295, 296f
Tenochtitlan, 104, 104f, 323f
teosinte, 101, 101f
Teotihuacan, 104, 323f
terraces
  in Andes, 106, 107f
  rice, 89, 89f
terra preta, 108–109, 109f
Thunderbird Glacier, 295f
Tiwanaku culture, 106
tree rings
  as climate proxy, 295, 296–297, 296f
  precipitation and, 296–297
  temperature and, 295, 296f
tropical Africa
  Bantu people in, 125–126, 126f, 127–128, 127f
  crop domestication in, 125, 126f
  early tools of, 124
  forest thinning in, 126
tundra, 264, 265f

Vavrus, Steve, 254
vegetation-albedo feedback, 264–265, 265f
virgin soil pandemic
    Amazonia and, 325, 325f
    Aztec Mexico and, 322–323, 323f
    CO_2 concentration changes during, 321f
    Crosby and, 320
    diseases associated with, 320
    evidence of, 321–322
    immunity and, 320–321
    Incan Peru and, 323–324, 324f
    North America and, 325–328, 326f, 327f
volcanic eruptions, 299, 299f, 300f

Wari culture, 106, 107f
weirs, 110
wetlands

Arctic, 24–25, 24f
    as CH_4 source, 20f, 176–177, 176f
    insolation and, 22–24, 23f
    monsoons and, 21–24, 22f, 23f
    north–south CH_4 concentration differences and, 25–27, 26f
wheat domestication, 55, 55f
windmills, 69, 69f
winter sea ice extent, 256, 257f
wobble, orbital, 7, 7f

yams, 124f, 125, 126f, 127, 130, 130t, 133f, 134
Yangtze River basin
    livestock and, 91–92
    rice irrigation, 91, 91f
Younger Dryas (cold climatic event), 19

Zebu cattle, 79, 81, 82f, 83f